To End the Arms Race

Books by the Author

Dynamic Principles of Mechanics, 1949
Nuclear Energy: Its Physics and Its Social Challenge, 1973
Wind Power and Other Energy Options, 1978

To End the Arms Race

Seeking a Safer Future

by David Rittenhouse Inglis

With a Foreword by Norman Cousins

Ann Arbor   The University of Michigan Press

Copyright © by The University of Michigan 1986
All rights reserved
Published in the United States of America by
The University of Michigan Press and simultaneously
in Rexdale, Canada, by John Wiley & Sons Canada, Limited
Manufactured in the United States of America

1989   1988   1987   1986   4   3   2   1

Library of Congress Cataloging-in-Publication Data

Inglis, David Rittenhouse, 1905–
   To end the arms race.

   Bibliography: p.
   1. Nuclear disarmement—Addresses, essays, lectures.
2. Nuclear arms control—Addresses, essays, lectures.
I. Title.
JX1974.7.154   1986   327.1'74   85-13938
ISBN 0-472-09367-3 (alk. paper)
ISBN 0-472-06367-7 (pbk. : alk. paper)

Foreword

David Inglis is a highly responsible theoretical physicist who, for the past few years, has been probing the implications on weaponry of the revolution in science and technology. He has been both consistent and persistent in his efforts to educate the American people—not just about the power of these new weapons, but about their threat to the human future. He is therefore eager to see strong public support for the establishment of governmental policies at home or abroad that promote approaches and policies for bringing the weapons under control and for strengthening the concept of workable world order. The prophetic nature of his analysis is readily apparent in his book and establishes his credentials for urging specific policies to meet the ongoing and deepening crisis. This chronological compilation provides us, therefore, with not only a political and technological record of the nuclear arms race, but a historic perspective from which to survey our present situation.

Looking back, Dr. Inglis identifies the origins of the deterrence theory, systematically dismantling the arguments that ignore the high level of overkill. He differentiates between the technical situation we faced in the early years of the nuclear arms race and the accelerated dangers today. Most importantly, he identifies the major roadblocks in arms reduction agreements. He has a realistic understanding of the complexities in dealing with the Soviet Union, but he also recognizes that these difficulties are no excuse for the Pentagon's persistent reluctance to pursue serious negotiations in arms reduction and control.

He astutely asks whether we can afford the luxury of delay in creating a genuine basis for national security. The record he has assembled is compelling and valulable and serves both to remind us of the ground already covered and to point the way to constructive and realistic openings. By documenting the course of the nuclear arms race, Dr. Inglis gives notice that we must harness the full measure of our accumulated resources and energies in the pursuit of control.

The reader is not left with a sense of helplessness or malaise—so often the case with analytical works on United States foreign policy—but with a sense of direction. Dr. Inglis is right in believing that the starting point for security is accurate information, clear perceptions, and a widespread and profound recognition of our common responsibility to safeguard life on this planet. We can be grateful that a scientist of the stature and conscience of Dr. Inglis has provided us with a design for a rational future.

Norman Cousins

Acknowledgments

Grateful acknowledgments are made to the following publishers and journals for use of copyrighted material:

Amherst (alumni magazine) for "Nuclear Arms Control Aspirations and Frustrations," *Amherst* 25 (Winter 1974).

Centennial Review for "A Specific Proposal for Balanced Disarmament and Atomic Control," and "Evolving Patterns of Nuclear Disarmament Proposals," *Centennial Review* 6 (Fall 1962).

Current History, Inc., for "Step-by-Step Disarmament," *Current History* 47 (August 1964).

Educational Foundation for Nuclear Science for the following articles from the *Bulletin of the Atomic Scientists:* "A Deal before Midnight?," volume 7 (1951); "The Stassen Appointment: Turning Point in Disarmament Thinking," volume 11 (1955); "Arms Control Effort Buried in State," volume 13 (1957); "Ban H-Bomb Tests and Favor the Defense," volume 10 (1954); "The Rest of the Test Ban," volume 19 (1963); "National Security with the Arms Race Limited," volume 12 (1956); "The Fourth-Country Problem: Let's Stop at Three," volume 15 (1959); "Nuclear Threats, ABM Systems, and Proliferation," volume 24 (1968); "We Haven't Really Tried," volume 11 (1955); "Shelters and the Chance of War," volume 18 (1962); "Conservative Judgments and Missile Madness," volume 24 (1968); "Freeze the Cruise," volume 40 (1984); "The Sweet Voice of Reason," volume 29 (1974); "Atomic Profits and the Question of Survival," volume 9 (1953); "Nuclear Energy and the Malthusian Dilemma," volume 27 (1971). Reprinted by permission of the *Bulletin of the Atomic Scientists,* a magazine of science and world affairs. Copyright © by the Educational Foundation for Nuclear Science, Chicago, Illinois 60637.

Foreign Policy Bulletin for "Urgent Need for Atomic Control," *Foreign Policy Bulletin* 32 (November 1952).

Heldref Publications for "Nuclear Energy: Rasmussen Revisited," *Environment* 18 (November 1976), p. 38; "Power from the Ocean Winds," *Environment* 20 (October

1978), p. 27. *Environment* is a publication of the Helen Dwight Reid Educational Foundation.

The Nation for "H-Bomb Control: Safeguarding the World," *The Nation* 179 (July 1954).

New Republic for "Why I Am for Stevenson," 135 (October 22, 1956); "Evasion of the H-Bomb Issue," 135 (November 5, 1956); "If We Just Go On Testing . . . ," 142 (April 25, 1960). Reprinted by permission of the *New Republic*. Copyright © 1956, 1960, The New Republic, Inc.

The Progressive, Inc., for "Energy Gluttony and Overkill," *Progressive* 36 (August 1972). Reprinted by permission from the *Progressive,* 409 East Main Street, Madison, Wisconsin 53703. Copyright © 1972 The Progressive, Inc.

Saturday Review magazine for "Antimissile Drag Race," *Saturday Review* 50 (February 1967); "H-Bombs in the Back Yard," *Saturday Review* 51 (December 1968); "Antiballistic Missile: A Dangerous Folly," *Saturday Review* 51 (September 1968). Copyright © 1967, 1968 *Saturday Review* magazine. Reprinted by permission.

World Federalist Association for "The Atom and Disarmament: Some Technical Aspects," *Federalist* 3 (May 1953).

Contents

Introduction

As science advanced from one wonderful discovery to another through the centuries, the time became ripe for my generation of physicists and chemists to discover nuclear energy and in particular to invent the atomic bomb. Each of us participating in the original bomb development had to decide at some point whether or not to do so. The bomb was meant at first as a deterrent, but after it was used as a weapon of war its further development took on a life of its own. The need for diplomatic initiatives to avoid the dangers of a perpetual arms race became a matter of public concern, and many scientists joined in the search for useful approaches. In the long contention between those opposing and those promoting the escalating arms race, there were more scientists opposing it, myself among them, than promoting it. My own motivation to participate in the original development of the bomb and in the opposition to its excessive deployment was doubtless influenced by my early upbringing and my later worldwide contacts. This book is mainly about the growth of ideas on why and how to curb the arms race, as reflected by some of my past writings on the subject, but by way of introduction we start here with mention of some of the earlier activities and associations that may have influenced my later attitudes.

I was born in a lucky generation, too young to be expected to face enemy bullets in World War I and old enough to serve in a specialized capacity far from battle in World War II. My father, likewise, was born near the close of the Civil War, and memories of that conflict were still fresh enough to emphasize the frequent occurrence of wars.

My first experience associated with war was as a war profiteer. I was eleven years old in 1916, and in Ann Arbor, Michigan, we lived near a small park surrounded by fraternity houses in which were billeted students taking military training as part of the war effort. My enterprising chum and I discovered that, when the hungry soldiers broke ranks from close order drill in the park, there was a market for handheld slices of apple pie. We walked downtown to buy large pies for twenty cents, cut them into six slices, and sold each for a nickel.

The war was far away, and affected home life very little. When it was over, the instruments of advanced technology that had been developed during the war came home, along with most of the men. We were excited to see real airplanes flying over town, those wooden biplanes doing spectacular loop-the-loops, perhaps piloted by some of the men who had drilled in the park, still drilling. My uncle, Paul Rittenhouse, came back as a lieutenant in the signal corps, and appreciated surplus radio equipment. He gave me a ruggedly constructed receiver with its galena crystal and cat-whisker contact nicely installed in a sturdy field box built for communications between trenches at the front. I made my first transmitter to go with it from a spark coil of a Model-T Ford, and learned Morse code to decipher the buzzing noises I could hear on the air. Advancing technology is full of surprises, and I had a great thrill when I first heard a voice on the air, not having appreciated that a mere crystal receiver could detect voice as well as code. It happened to be the voice of my assistant scoutmaster using a surplus field transmitter, containing real vacuum tubes, at the university station a mile away. There was always something to learn in ham radio that I pursued during my high school days. War surplus was thus an introduction to my lifelong curiosity about the workings of nature. Physics came naturally as an important emphasis in my education and later interests.

But there were also influences turning me toward awareness of the world's peoples and their problems. Perhaps the first was my Aunt Agnes, an advanced-thinking socialist and, as such, a thorn in the flesh of her loving and caring conservative-businessman brothers, my father and Uncle Jim. She induced me to bring along a few of the boys of my age, ten or eleven years old or so, to form a club that would meet at her house one afternoon a week. The arrangement was that she would provide the materials for whatever hobby we wanted to pursue, making model airplanes or bird houses or such, provided we would listen to her for fifteen minutes at each meeting. Our thoughts were turned gently toward peoples' needs and the inequities in the distribution of goods and opportunities worldwide.

There were of course many other influences outside those of a healthy home, such as scouting when the Boy Scout organization was young and fresh, summer camps, and the Hi-Y Club, a YMCA-sponsored high school activity with a moral tone something like Sunday school. I was president of it, and my one experience at running for political office, unsuccessfully, was for state president of Hi-Y. There was also the Foreign-American Club, of which I was president, too. Of my high school teachers, the one who influenced me most was a history and civics teacher, Edith Hoyle, who injected human compassion into history. Ann Arbor High had many foreign students who came to this country to enter the university but found it expedient to start with the local high school instead. Miss Hoyle sponsored the Foreign-American Club. We had friends from many lands. Inter–high school debating also developed my interests in national and international affairs.

A couple of years later, when I was a sophomore at Amherst College, I again felt the influence of a fine history teacher, Professor Lawrence Packard. The subject of his special study and of the course he taught was ''The Origins of the World War'' (not yet known as World War I). On this subject I had until then not progressed far beyond my boyhood wartime view of the demon enemy, the lust of Kaiser Bill for world conquest, and the necessity of our sending troops to Europe in defense of America to help defeat him there, so he could not conquer first Europe, then Mexico, and then attack us from the Mexican border. I was amazed and intrigued to learn from Professor Packard how the gradual buildup of tensions accompanied by the determination of the large states to stand firm by their commitments to client states, and the quarrelsomeness of the client states emboldened by this backing, had ignited the war that nobody wanted. Although this could be seen as just a continuation of the cruelty of history on a grander scale, it seemed utterly stupid on the part of modern mankind that the wholesale carnage of trench warfare, then still fresh in mind, with young men swarming over the top to be mowed down by machine gun fire, should be triggered by the bungling machinations of diplomats competing for national prestige and influence, and able to rally their populations behind them with exaggerated perceptions and hollow slogans.

I next had the good fortune to spend my junior college year traveling around the world as a student in a unique educational experiment known as the University Afloat. Studies on shipboard included history of the lands we were about to visit, and shore trips afforded contacts with local students. There was the fun and fascination of foreign travel while my eyes were opened to the great cultural and economic differences between the East and the West. We saw the benefits and sufferings of colonialism as the result of past wars but all was peaceful, except that an ongoing local revolution prevented our visit to Peking.

In the Philippines my brother and I had a close look at a sample of the worldwide projection of American military power. Mother's cousin was the admiral in command of the Cavite Navy Base across the bay from Manila. While visiting there with him and his family we attended a festive army-navy ball in Manila that lasted late into the night as scores came in by wireless from the afternoon army-navy football game halfway around the globe. The rivalry seemed so intense, at times so bitter, that one would hardly suspect that the two services were parts of the same national team. I sometimes recall that impression as I see the nuclear arms race spurred by competition between the services for pieces of the budget, where a juicy item if given to one must be given to each.

On our old coal-burning ship we sailed right through the impressive crater of Krakatoa, whose eruption about fifty years earlier had caused a global year without a summer. We may now look back on that awesome event as a mild preview of the global deep freeze that could result from the smoke of many fires set by a nuclear war. Pulverized rock in the stratosphere from that eruption scattered sunlight, letting some of it filter through, whereas so much smoke up there would absorb the sunlight and send it back into space as heat.

When, at the end of the European part of our travels, the ship returned to New York in early spring, I remained in Europe for the spring and summer. I first toured Great Britain with my parents, then lived with a retired French journalist and his wife in a Left Bank apartment in Paris, practicing French, and then as a foreign summer student at Heidelberg, learning German. I worked my way back to New York as a waiter in the third class of a trans-atlantic liner.

For the next dozen years or so thoughts about war faded into the back-ground as I lived through the adventure of becoming a scientist and settling into rather normal academic life. I started my career as a theoretical atomic physicist, using the miraculous techniques of the new quantum mechanics to explore some of the detailed aspects of atomic structure that presented them-selves. After the discovery of the neutron in 1933 it became apparent that there were similar manifestations of structure on a scale a thousand times smaller within the atomic nucleus and I became primarily a nuclear physicist, applying quite similar techniques again there. I even went into experimental nuclear physics for a while when some of the data I wanted in theory were lacking. There was a great satisfaction in taking one's own small part in forwarding the advancement of science, delving into the secrets of nature. There was a feeling, too, that this was useful in the long run, increasing man's mastery over nature, though each step was motivated mainly by scientific curiosity.

During those dozen years of the late twenties and the thirties I had many occasions to develop an international viewpoint as is almost inevitable for a scientist, becoming acquainted with foreign lands and people. International scientific meetings are a dime a dozen nowadays but back in the twenties and thirties the Summer Symposium on Theoretical Physics at Ann Arbor was something special. Right there at home, as I was a graduate student and thereafter, I first made the acquaintance of most of the world's foremost theoretical physicists, many of them from Europe.

After I taught for a year at Ohio State, I took a leave of absence in 1932 to spend a year in Europe, parsimoniously stretching my first-year salary to cover two years, for fellowships were hard to come by during the Depression. That was a year of many contacts, for I was free to move around from one physics institute to another, with visits to friends and mountains in between. I was working mainly on one of the many subjects on which Heisenberg had made important advances, ferromagnetism, and I planned to be with his group at the University of Leipzig for the academic year, starting in the late fall. In the summer semester the theoretical physics group at Goettingen under Max Born was a lively one. I hit it off particularly well with a young privatdozent, Edward Teller, the beginning of a long and rewarding friendship, though our political views have since diverged drastically as he has been contributing so effectively to the arms race. Our present political differences seem to reflect very different early influences, for he had grown up amidst old European rivalries and as a boy had experienced Soviet aggression in Hungary.

Another friend at Goettingen was Walter Heitler, later professor at Dublin, with whom I went mountain climbing later in the summer from a tiny Tyrolean village visited by his family. From Tyrol I went to Rome, where the fall semester started a month earlier than in Leipzig, to be with Fermi and his group. Rome is of course a pleasant place to be under almost any circumstances and to be there mainly to talk physics but with time to look around was especially pleasant. I ate in restaurants sometimes with Ettore Majorana, a brilliant young Italian physicist who mysteriously disappeared a few years later, and more often with my American friend, Eugene Feenberg. This was the time when Mussolini's fascism had taken over the country with a strong-arm regime and was beginning its military adventurism in Africa to distract attention from economic troubles at home. There was a repressive spirit in the air. Eugene and I felt that we would be watched with suspicion if we were overheard mentioning Mussolini in our derogatory discussion and we used the code word *Al Capone* instead.

Fermi's little institute in modest old quarters was then still engaged mainly in fundamental theoretical developments, such as his early formulation of quantum electrodynamics that I had heard him lecture about in Ann Arbor the previous summer. Not until a year later, after the discovery of the neutron, did it become a beehive of experimental investigation that barely missed discovering the fission process. Fermi and his collaborators actually observed fission product elements but misidentified them as new elements heavier than uranium. Fission was not discovered until six years later. I have since wondered, in thinking back on that little laboratory, whether some benign Providence was postponing the discovery of fission until the fascist powers were too preoccupied with the war effort to develop an atomic bomb.

Enrico Fermi, Werner Heisenberg, and Wolfgang Pauli, with each of whom I had pleasant contacts during this year abroad, were three of perhaps half a dozen most important giants in theoretical physics who had ushered in the new age of quantum mechanics just a few years before, in the mid-twenties. But on this occasion Fermi turned important experimental physicist as well. A decade later he combined his theoretical and experimental prowess to lead the production of the first nuclear chain reaction at Chicago.

In Leipzig I found Heisenberg's theoretical physics group entirely congenial, enlivened not only by his personality but by his very young assistant, Weitsaecker (Heisenberg was all of thirty by then), and an important visitor, Felix Bloch. Serious work was interrupted by weekly social evenings featuring Ping-Pong and rapid chess, and by Tuesday morning walks down to the Gewandhaus to hear the formal orchestra rehearsal led by Bruno Walter. I also attended late-afternoon recitals of Bach organ music at the Thomaskirke where it was written. When Heisenberg graciously entertained me and two others at his bachelor apartment, after dinner served by his housekeeper, he delighted us by beautifully playing classical piano music.

This intellectual and cultural life that I experienced was a fine manifestation of the recovery of Germany from the traumatic experience of defeat in World War I. The runaway inflation that wreaked economic havoc was past, the economy was struggling in recovery and seemed quite healthy, while living was still cheap in American dollars. The psychological reaction to the defeat in war led to political turmoil. While the Germans I knew, mostly intellectuals, watched seemingly helpless with foreboding, there was a rise of aggressive nationalism harboring a spirit of revenge for the Versailles treaty that had imposed the conditions of the defeat. In a sense, though I could not be sure of it at the time, I watched a new war in the making. I could see how the most narrowly nationalistic elements of a powerful people could force the rise of a dangerous demagogue despite the misgivings of more thoughtful citizens.

During my first visit to Germany in 1927 the Nazi party was young and small and I had seen nothing of it at Heidelberg, but I had noticed one evidence of resurging nationalism. I had wanted to learn glider flying and there was a famous German gliding school in the dunes on the shore of the Baltic. I obtained the brochure and filled out my application for a session just after the summer semester at Heidelberg, but then noticed in the fine print that the cost was only one-third as much for German citizens as for foreigners. This was my first realization that Germany was starting to get in shape to fight again. The Treaty of Versailles forbade aircraft pilot training for Germans, but glider piloting was clearly being used as a preliminary substitute. By way of protest, I decided not to join and subsidize that course. Sailplanes have changed a lot since those early wooden biplanes that the students hauled back up the dunes for takeoff, but I still haven't learned to fly.

By the summer of 1932 when I was at Goettingen the Nazis had grown strong and threatening. Everyone I knew in the academic community was against them and worried, but there seemed little they could do about the growing menace, which some took seriously and some did not. A huge mass rally was organized by the Nazis in a park at the edge of town and a group of us out of curiosity attended to hear Hitler speak. We stood long in the crowd before his plane finally arrived. He mounted the podium and harangued menacingly for fifteen minutes and then as dramatically departed. By then I could understand most of an ordinary German conversation, but understood only a little of his harsh staccato. I could not but feel that we had done wrong to attend and lend our presence to the appearance of strength of the movement.

By fall and winter when I was at Leipzig the show had intensified. Young men in uniform or with swastika armbands paraded and strutted. I occasionally heard shots down the street and did not know what to make of them. But still the academic life at the university continued serenely, with increasing anxiety but little affected.

The winter semester ended a week earlier in Leipzig than in Munich and I spent that week visiting Sommerfeld's institute at Munich. On the fifth of March there was an important election there in Bavaria. Sommerfeld invited me along with most of the young people of his institute for a social gathering in his home that evening. Most of us were not so much involved in politics but that we could make a pleasant occasion of it but the wise old Geheimrat professor himself, Sommerfeld, was very melancholy. He understood that the unfavorable vote that day, the memorable Beyerischer Wahl, was the beginning of the end. Not long thereafter Hitler became Reichskancellor.

While in Munich I met some skiers and joined the University of Munich ski club for the sake of a spring vacation skiing excursion in Tyrol. This was my third expedition into the Tyrolean Alps that year, for I had spent two weeks of Christmas–New Year's vacation in a little hotel high up on the mountainside above Kitsbuhel skiing with three other theoretical physicists, Rudolph Peierls, now Sir Rudolph, Viki Weisskopf, whose career on the forefront of physics has since continued past being head of the great all-European physics laboratory at Geneva and of the Physics Department at MIT on into the new realm of the quark structure of nucleons, and Max Delbruck, who later received his Nobel Prize not in physics but biology. (They have also become outspoken opponents of the arms race.) The ski week with the Munich students in the spring, this time centered at a cluster of high chalets above the Ötztal from which we climbed the highest mountain in Tyrol, was the only one of my three alpine excursions without other physicists, but each of them gave me a feeling for the exhilarating glories of the Alps as well as the variety of life and problems of Europe.

I next arrived in Zürich with my arm in a sling, for not all had gone well with the skiing, intending to visit briefly with Pauli and his group at the Swiss Federal Technical Institute, the E.T.H., on my way back to Leipzig. There I found Felix Bloch, recently arrived from Leipzig. He advised me, "Dave, don't go back to Leipzig. One can't study there any more. The Nazis make too much commotion." So I didn't, and instead spent a very pleasant and quite productive spring semester at Zurich. Not only were there the promising young postdoctoral physicists Casimir, as Pauli's assistant, Bloch, Bhaba, and Elsasser with often-jovial Pauli at the E.T.H., but also meticulous Gregor Wentzel as professor at the university down the street, most of them friends of mine on into future years. Casimir was already a good friend, having come to Ann Arbor with his professor, Kramers, from Holland a previous summer, and we were pals in Zurich that spring. I became fascinated with a puzzling experimental result that Pauli pointed out to me on a nuclear aspect of atomic spectroscopy observed at the university there and spent much of my time figuring out an explanation of it.

After a good spring in Switzerland, away from any local concern about the Nazis, I returned to my job at Ohio State to teach in the summer quarter. There was worsening news about people losing their jobs in the lean year of 1933, and I did not feel confident enough about mine to stay away for another summer in Europe. On the way to Hamburg for sailing to New York I stopped off at Leipzig to pick up the belongings I had left behind there. There I saw increased evidence of the Nazi nastiness, particularly in the form of rough lettering in yellow paint on many store fronts, *Judengeshäft*. Some were still occupied, some closed. Misdirected mass movements, in order to prosper, seem to need an enemy to hate.

The following year I was married and by three years later, in 1937, Betty and I had saved up enough to spend the summer in Europe. In those days the Nazi threat was gradually growing stronger, with transgressions of the Treaty and the beginnings of armaments, and the thing to do in protest was to "boycott Hitler." We had done a lot of paddling down American rivers in a demountable, rubber-hulled folding kayak of a type that was then most popular in Germany, where it was known as a *Faltboote*. It was an old one that I had bought secondhand in Hamburg. We bought a new one to replace it in Paris, just after landing, selecting a good patriotic French brand of *canöe pliant,* rather than buying it in Germany where they were cheaper. We left it in Paris to be delivered later, and after spending a couple of months, first in Rome where we saw Amaldi and others (though Fermi was absent from his modern new institute), and at Zurich where we saw something of Pauli and Wentzel, we had it delivered to a village railroad station on the French side of the Swiss border near Geneva, so as to avoid customs on it. As we unpacked the package there on the banks of the Rhone, preparatory to riding those melted glacier waters far downstream, we noticed first that the little wooden parts of the frame were labeled, in French, "Made in Czechoslovakia." But when we came to the main part, the rubber-and-canvas hull and deck section, it was labeled "Made in Germany." We were later glad to have chosen that year for our splendid European summer. The next was too close for comfort to the outbreak of war.

That summer I arranged by cable to accept an invitation to teach at Princeton, transferring from Pitt, and by the time this country entered the war, in 1941, I was teaching and doing some experimental as well as my usual theoretical physics research, mostly nuclear physics, at Johns Hopkins. Our physics department there became associated with the Carnegie Institution of Washington, under the wartime Office of Scientific Research and Development, in organizing the effort to develop a proximity fuse for antiaircraft artillery shells. I first became involved in war work in that effort, making calculations to help maximize effectiveness, while still doing half-time teaching. The war work became a full-time commitment with the Army Corps of Engineers at Aberdeen Proving Ground a year later.

There was a tug of conscience in getting drawn into war work. Everything I had learned since boyhood led me to abhor the idea of war as a way of settling international quarrels. I had felt the seriousness of the way Hitler's trickery and hate campaign could turn the minds of a great people toward war and I was amazed and alarmed when governments did nothing in response to his occupation of the Saar, when a relatively modest show of force might have prevented the most serious phase of German rearmament. Once we were drawn into full-scale war across both the Atlantic and the Pacific, provoked by raw aggression, there was little doubt that one should work toward improving the effectiveness of our forces to hasten the end of the carnage. This has doubtless been a classic motivation for participation on both sides of most wars.

During those years 1939–43 I knew about nuclear fission and the theoretical possibility of a chain reaction but had only the vaguest idea that something was being done to try to exploit it. I had been among the first to learn about the fission process when Nils Bohr, on a visit from Denmark in early 1939, after landing in New York brought the news about fission directly to a little meeting on theoretical physics at the Carnegie Institution Laboratory in Rock Creek Park in Washington. Just before leaving he had learned about fission from Otto Frisch, who had just reached the crucial insight in discussions with his aunt, Lisa Meitner, who had collaborated in the original experiment in Germany. For the next year or so there were articles about fission in the *Physical Review* but these suddenly stopped as the physicists involved imposed a voluntary censorship, thinking the process might be useful in war. From my eminent senior colleague at Hopkins, professor R. W. Wood, who had wide connections in the physics world and who could not repress his enthusiasm for anything that interested him, I heard hush-hush rumors that a chain reaction had been achieved in great secrecy. This in a reactor would not require massive separation of isotopes, but an explosive chain reaction in a uranium bomb would. I was much impressed by the great pains that were taken back in the thirties to separate minute quantities of isotopes. We were accustomed to things being done on what now seems like a very modest scale indeed in scientific laboratories before the war and it seemed to me implausible that kilogram quantities of these hitherto extremely rare special materials, as would be required to make a bomb, could be prepared during the course of the war, if ever. This skepticism, which was not mine alone, may stand as tribute to those brave physicists and chemists who had the imagination and initiative to undertake the unprecedented effort that eventually produced the needed materials. That undertaking, too, was carried out in secrecy, so there were surprises in store for me when I joined the bomb effort.

1 Need for Disarmament Planning

A laboratory to invent and construct an atomic bomb was established at Los Alamos, New Mexico, only four years after the fission process, fundamental to nuclear energy, had been discovered in the course of pure research in Nazi Germany near the start of World War II. When I was asked to join the nascent project in early 1943 I first learned that the prospect for the military use of fission was considered so promising that isotope-separation plants were actually being constructed in secret, enormously larger than any prewar scientific endeavors. There were still doubts. Nuclear cross sections, for example, needed to be measured at the new laboratory to learn whether nature had indeed put in men's hands the material to make an atomic bomb. While such a weapon could quickly end the war, it was clear that it would be no boon to mankind in the long run. There was a question whether one should participate in the endeavor. Such large projects had never been possible in a democracy in peacetime but apparently could be in wartime. In a totalitarian dictatorship such effort could presumably also be marshaled after the war. If such a weapon could be made, it therefore seemed clear that it eventually would be and imperative that it should first be achieved in a democracy that would not try to conquer the world with it and could take measures to avoid its misuse. With faith in the workings of democracy I could thus see the project as a worthy cause, not only for the immediate objective of preventing Hitler from having the bomb first during the war but also for the long-term good of the world.

After the bomb was achieved and the war ended in 1945, I was one of many scientists on the project quick to return to the pursuit of pure science and teaching that had been interrupted for the duration of the war emergency. When the existence of the bomb was made generally known, it seemed to us

11

who had for some time seen it coming that the public and officialdom were slow to comprehend its significance for world affairs, and that it was up to us to inform and convince them. Some devoted a great deal of time to this in Washington, giving birth to the Atomic Energy Commission (AEC) and the Acheson-Lilienthal-Baruch plan that became the basis for U.S. arms control negotiating policy. Many of us felt some confidence that, once people were informed of the nuclear dangers, responsible officials would take reasonable steps to mitigate the threat. Just as it was faith in democracy that led to my getting into the arms development project in the first place, it was through a similar faith that I confined my modest political activities mainly to informing the Baltimore public about the nuclear threat during the three years I remained at Johns Hopkins after the war.

When I transferred my pursuit of pure science to the Argonne National Laboratory in 1949, I encountered there as head of the computer division Donald Flanders, a good friend who had served in a similar position at Los Alamos. Our almost daily luncheon conversations turned to the problem of avoiding an endless nuclear arms race now that the Soviets, too, had achieved the A-bomb and the advent of the H-bomb was anticipated. Arms control negotiations had become frozen, with the Soviets repeatedly proposing general and complete disarmament that they knew we would not accept, and with us proposing the Baruch plan for internationalization of atomic efforts that we knew by then the Soviets would not accept. Dr. Flanders and I were rather shocked as we came to appreciate in our discussions in those days of much secrecy that there was no sense of urgency about nuclear weapons in official circles to do anything but build more of them. Everyone in government was busy with something else and there was no inside source of ideas on arms control. When we put some of our concerns together in the first of the following articles, we suggested among other things that there should be an agency in the government with responsibility for developing ideas about arms control initiatives. This seems to have been the first published proposal of this sort. Ten years after publication of the article, the U.S. Arms Control and Disarmament Agency was established, having as just one of its functions the development of arms control initiatives.

One important part of the effort of scientists to inform the public about the problems and dangers of nuclear war was the establishment of the *Bulletin of the Atomic Scientists*. Its name, by the way, has been preserved from the first issues just at the end of the war when it was indeed a small bulletin circulated mainly among scientists. It is intended to be written by scientists for the general public but too many of the public have ignored it under the impression that it is by scientists for scientists. The conviction among these scientists in the late forties was that something must be done very soon to avoid an arms race that would end in world catastrophe. This sense of urgency was expressed by the *Bulletin*'s clock with its hands approaching midnight. Hence the title of that first of these articles, "A Deal Before Midnight?" As officialdom has failed year after year to grasp successive opportunities to initiate more than token restraint, the conviction has persisted and the message of the clock remains relevant.

In those years there were occasional international conferences to consider the matter of nuclear arms limitations. Each time the government found itself without an adequate policy on arms limitation and each time a committee consisting of prominent individuals, most of whom had little relevant background for the crash study they undertook, was appointed only a few weeks in advance of the conference to put together a credible negotiation stance. The first step beyond the use of this temporary expedient was President Eisenhower's appointment of former governor Harold Stassen as the president's special assistant for disarmament with the rank of ambassador. Negotiations as a result of studies under his leadership were coming closer than some would have liked to achieving some meaningful arms control agreement with the Soviets when his mission was suddenly withdrawn after AEC chairman Lewis Strauss dramatically presented to President Eisenhower Drs. Teller and Lawrence, who brought enthusiastic tidings about the possibility of developing a "clean bomb," a powerful H-bomb with substantially reduced production of radioactivity. The urge to develop a new weapon took precedence over the search for restraint in the arms race.

These were years when public and official attitudes toward the arms race were being formed that have persisted ever since. After having a head start we felt confidence in our ability to stay way ahead in the race. The number of bombs, still to be delivered by aircraft, was increasing from tens to hundreds to thousands, though the actual number, and everything that might help guess it, was for too long a time kept secret. The belief in Soviet expansionism, to be contained by military strength, was first firmly established in the public mind in the time of Stalin and persisted. Some of us agreed with the need for military containment but recognized the danger of an open-ended arms race and felt the urgency of exploiting a unique opportunity to stop the arms race by negotiation while the nuclear stockpiles were still relatively small. Writings to this effect were mostly lost like a voice in the wilderness. Some of them reflecting the mood of the time are included in this chapter.

As we look back through these papers on those earlier years of the arms race, we see several ways that the problem of controlling nuclear developments was different then from now. Yet then just as now there was a confrontation between those who seek national security through building even more nuclear weapons systems and those who seek to avoid an even more dangerous situation in the future by ending the arms race and eventually reversing it. Then as in more recent years the arms promoters have held sway. They have permitted negotiations to proceed cautiously far enough to give the public the impression that reasonable efforts are being made to negotiate restraint, but when the negotiations have seemed close to success they have been terminated. The termination was sometimes deliberate and sometimes triggered by an unfortunately timed incident, such as the shooting down of the very secret U-2 reconnaissance plane over Soviet territory just before a crucial meeting late in the Eisenhower administration. The negotiations seem to have been window dressing in the eyes of the arms promoters, needed for public relations. The way they were permitted to proceed, sometimes only as a sham, is indicated in a revealing quotation in the fourth of the following articles. Yet they were pursued with diligence and sincerity by those favoring a negotiated end to the arms race. President Eisenhower as a military man seems to have been torn between the two views, between his genuine desire to end the arms race, and his susceptibility to arguments for developing new weapons. As a commanding general he had made a practice of establishing a special group of his officers to think about long-range planning, a sort of "think tank," and it was in this same spirit that as president he appointed Stassen, an ardent supporter of arms limitation, as his special ambassador for disarmament.

The Stassen appointment, when it was made, of course raised hopes. As suggested in that fourth of the following articles, it seemed that it might fill the need cited in the first article for a think tank on disarmament methodology. There were similar hopes some five years later when the U.S. Arms Control and Disarmament Agency was established. Actually its responsibility for current negotiations absorbed most of its resources and seemed to leave it with little interest in thinking out new schemes for disarmament.

The technical situation then differed from what we face now in two important ways: first, the number of possible nuclear bombs, or the amount of special nuclear material that had been produced to make them, was still small enough that there was some hope of accounting for most of it by inspection of the production plants, and second, delivery was to be by slow and vulnerable bomber rather than by intercontinental missile. While the amount of prospective destruction by nuclear attack was enormously greater than with nonnuclear weapons, it still did not represent multiple overkill and there was a valid military meaning to fairly small changes in the numbers of bombs available, which is no longer the case.

Particularly because of the possibility of a breakthrough in the defense against bombers, it did not seem as preposterous then as it does today to assume that the Soviets might see some prospective gain in launching a surprise nuclear attack. In my discussions I went along with this usual assumption.

This made it even more important then than it is now for any arms limitation agreement to provide verifiability of detailed compliance. The problem was then, even more essentially than now, made difficult by the Soviet penchant for excessive secrecy. Not only did promoters of increased armaments seem to welcome such obstacles to disarmament, but many others accepted them as making efforts toward arms control hopeless. Many of us, however, took the difficulties as a challenge. It seemed so clearly in the best interest of both sides to avoid the dangers of a perpetual arms race that we felt it must be possible to devise an arms limitation scheme that would overcome the difficulties and would appeal to both sides in spite of all the suspicion, ill will, and ideological difference. There was a chance, at least, that what was lacking was the right idea, and the chance seemed worth pursuing. The papers in this chapter are concerned with setting up an official search for the right idea. Those in chapter 5 are representative of a private search.

Back when most of these papers were written, I felt that the international political difficulties were the main obstacle to achieving an arms limitation agreement and that finding the right scheme to overcome them might lead to a national will to negotiate effectively. In retrospect I believe that we were then already well on our way to a situation we can see more clearly now. The problem is not so much the lack of ideas on how and what to negotiate. The main obstacle to making a sincere, high-priority effort lies instead in our national decision-making structure, dominated by pressures from vested commercial and military interests exerted on officials elected with too little concern for their views on foreign policy by an electorate swayed mainly by pocketbook issues. The economic and public relations fallout from enormous expenditures on excessive nuclear weaponry molds the attitudes of officials and citizenry alike. The arms race has thus become a colossal establishment in American politics. Dedicated individuals and organizations opposing it have financial resources so much smaller, in a contest where money counts, that their confronting it seems like David facing Goliath. Thirty years of opposition have by and large failed, for the arms race continues. Yet avoiding nuclear war, first by negotiating an end to the arms race, is a righteous cause and with perseverance it could prevail, though the time is very late. There still is a need for improved ideas on disarmament planning.

A Deal before Midnight?

[1951]

During the first few years of the atomic age there has been a dimming of the early hope that international agreement might prevent the calamity of an atomic holocaust. But even as we struggle to postpone the evil day by maintaining the upper hand in the conflict between East and West, we should be making a vigorous exploration for further ideas about possible agreements.

For the unlimited growth of stockpiles is progressively adding to the difficulties of both the political *and* technical problems of atomic control.

A-Bombs Liability to Both Sides

Here in the West we have come to feel that we need the atomic bomb. It has served us well in the short period of our "monopoly." Not only did it bring the end of World War II, it is probably responsible for the freedom of Western Europe from Soviet domination today, or at least it permitted us the luxury of bringing our troops home in response to public demand while leaving a vacuum of conventional power in Europe. It further made possible military economies at a time when unusual expenditures were needed for construction and reconstruction.

Today two very different parts of the world face each other with distrust and enmity, each possessed of atomic bombs; yet for neither of them does the atomic bomb appear to hold an advantage for future plans commensurate with the dangers involved. The main aim of the Western peoples is to make their part of the world safe for their way of life, and this necessitates defense against aggression. The role of atomic weapons in this defense is chiefly the threat of retaliation. But the stability of this position is very precarious indeed. It takes only one fool in power, one jittery decision, to pull the trigger and unleash unprecedented destruction.

In the Soviet-dominated part of the world an important tenet is that communism and capitalism cannot coexist, so the latter must be destroyed. One might conclude that the atomic bomb would ideally serve the Soviet purpose—destroy the rest of the world and let the Soviet part of the world prosper alone, even if it must rise from the ashes to do it.

Coauthored with Donald A. Flanders

17

But the Soviet leaders are probably not that ruthless. They doubtless aspire to dominate a world that is more nearly intact. Thus they might use atomic weapons in a war of nerves, in which they would hope that the iron nerves of the few in the Kremlin would triumph over the timidity of the many in the despised and divided capitalist countries. The growing strength and unity of the Western nations may well convince the Soviet leaders that this course will accomplish little except to keep the world sitting on a powder keg, with the grave risk that the powder will be ignited, thus destroying the prize they seek.

Lack of Long-Term Policy

The rejection of the Baruch plan and the frustration arising from Soviet political methods have given us a feeling that there is little hope of avoiding an eventual atomic world war. Needled by aggression in Korea and elsewhere we are proceeding with a show of confidence to try to postpone the threat of atomic war and to prepare ourselves, at least psychologically, to survive it as some sort of society if it comes.

Sometimes it would seem that the seriousness of the situation was being belittled, in that we are being told under brave titles how to survive atomic attack—essentially by being lucky enough to be far away from where the bombs hit. Nevertheless this may help steady our nerves against future Soviet rattling of the sword. More significantly, we are attempting to unify the free world, preparing it for nonatomic resistance to nonatomic aggression and stepping up our atomic program.

Our overall policy seems to be to devote the remaining time within which we may be supposed to have a clear superiority in atomic armaments to building up within the Western powers military forces better able to cope with the more conventional types of forces in which the Soviets may be assumed to have superiority.

We hope that when this object has been achieved the Soviets will find it to their interest to make and keep agreements. In the domain of atomic energy, however, we seem to have no plans, except possibly to re-offer the proposals already made, for the acceptance of which our main reliance would seem to lie in the possibility of internal change within the Soviet Union. The effectiveness of terrorist government discourages the hope of a popular uprising. It seems unlikely also that there will be a softening of Soviet foreign policy as a result of Stalin's death. There may be a palace revolution, but there do not seem to be the makings of a tolerant government within the present Politburo. The chief hope would seem to lie in a gradual mellowing of the Soviet government with growing maturity. This is quite conceivable in the course of a half-century or so, if the atomic crisis can be held off that long.

Atomic Control Increasingly Difficult

The overwhelming weakness of this policy arises from the fact that atomic energy becomes increasingly difficult to bring under control as time passes. No matter what sort of control scheme is devised, it is going to have to provide some assurance to each side that the other is not concealing a stock-pile of atomic explosives. Near the beginning of the atomic age, before there had been much time for the few existing plants to produce much fissionable material, this would have been relatively easy, although even then unhampered access to and inspection of the production plants would have been necessary. This was the principal concern of the Acheson-Lilienthal report.

But now that production has been under way on both sides for some time there is the added problem of estimating total past production. Clearly this can only be done with limited accuracy, which we might assume to be represented by some fixed percentage of the total past production. This means that when the total stockpile of either side is so large that it would be possible to conceal an amount sufficient to have a decisive or important effect in a war, it may no longer be possible to initiate a control scheme. For it would then be practically impossible to write a disarmament treaty that would not require at least one side to take an enormous risk that the other side might be preparing an effective surprise attack.

Thus there is at some time, perhaps in the fairly near future, a point of no return beyond which the world will be doomed to the threat of atomic war until that war takes place, unless by some political freak one side is able to gain the adherence of the other without the use of atomic weapons.

Possibility of Limited Interim Control Schemes

An ideal atomic control scheme would eliminate competing stockpiles entirely, as the Baruch proposal aimed to do. Now that competing stockpiles exist it may be too late to seek such an ideal solution except as the result of a long evolution. A first step in that direction might be an agreement that tolerates the existence of opposing stockpiles limited to a size well below saturation. The situation would still be dangerous, but less so than with unlimitedly increasing stockpiles.

Today's is an especially distrustful world, and no atomic control scheme is worth discussing that will not operate in spite of mutual distrust. Under these circumstances the possession of fissionable materials by both sides rather than by just one may be advantageous in initiating an agreement. There is serious uncertainty in the ratio 0/0. But if the sizes of the opposing stockpiles are not too disparate (and do not greatly exceed the ultimate saturation), there may be less danger to either side if an agreement is made that permits each side to keep a declared amount in comparison with which the error in estimation of the actual amount is not decisive. Thus it seems that serious consideration should be given to the technical and political feasibility of such an approach, among others.

Temporary Favorable Situation

In this connection there is another factor that would seem to be favorable to success in spite of past failure. Whether an agreement to limit atomic armaments is to be reached independently of agreements on conventional weapons or not, considerations of the latter must certainly enter into negotiations on the former. Thus the rearming of the West may for a time contribute to the possibility of reaching agreement, since atomic armaments will no longer be regarded as our sole defense.

Even this relatively favorable situation may last for only a few years if, in spite of the general technical considerations concerning the critical mass required for a chain reaction, the future tactical uses of atomic field weapons may be as important as suggested by the remarks of Senator Brien MacMahon, which appear elsewhere in this issue.

Our experience since the close of World War II has shown us the difficulties of obtaining Soviet adherence to agreements in situations where we were unable to enforce compliance. The atomic control problem is so tough that we should now be vigorously attempting to formulate the nature of the agreement we shall seek when we shall have had some success in our policy of creating a situation in which the Soviets will make and keep agreements. *The effort expended in strengthening our arms is enormous. We should not stint the effort needed to define our aims.*

Intensive Search for Ideas Needed

The prospect of atomic destruction is indeed much too grim to permit us to stop negotiating, or even to stop thinking about the problem, after trying with only one scheme. Though the aims of that scheme were about the best that we could reasonably strive for, there are other less far-reaching aims which, if achieved, would greatly reduce the threat. Perhaps the minimum aim worth striving for is to limit the growth of stockpiles to a point well below saturation, which would have the twofold effect of keeping the temptation for atomic surprise attack within bounds and of maintaining the possibility of eventually establishing fully effective control.

It is perhaps natural for the man in the street to think that our government has all these matters under adequate consideration, so that he need not worry about them. But the past failures of governments to anticipate their needs provide precedents that lead one to suspect that not all stones have been left unturned. We may assume that the president and secretary of state are giving serious attention to the problem, but we must remember that they are intensely preoccupied with a great variety of other problems.

Their position in diplomacy is like that of a commanding general in the military field. It is well known that in World War II generals and admirals and their staffs, who were preoccupied with specific problems, received inestimable aid from "operations research" teams. These were semi-independent groups of technically trained workers who studied the overall technical problems within the framework of their combat applications, and in some cases achieved astounding success by revolutionizing the use of weapons. In similar fashion in the diplomatic field the president and his staff should be supplied with creative ideas by a special group of imaginative thinkers dedicated to the single problem of atomic control in its relation to our whole diplomacy.

Such a group should be unhampered by secrecy restrictions and should be free to solicit aid from any relevant government agency. It should be charged with the task of assessing the assurances and risks, the advantages and disadvantages, of various control schemes that have been proposed, but especially with devising new schemes that might have better chances of political success. The fact that isolated thinkers preoccupied with other problems have failed to turn up better schemes does not mean that they cannot be found.

The group must clearly have not only scientifically and technically trained members but also men versed in politics, statecraft, and perhaps also mass psychology and other specialties, since any technically sound result must be put into a form that has the best prospect of acceptance in spite of world tensions, and the framing of prospective agreements will require attempts to maximize the desirability to both sides, perhaps even balancing concessions inside and outside the field of atomic energy, and attempts to minimize the economic sacrifices and military risks involved.

The fact that the group would have direct responsibility to the highest administrative authority should make it possible for the president to obtain the full-time services of creative men of the highest calibre.

The Acheson-Lilienthal Committee gave an example of the imaginative approach and mutual stimulation that such a group can achieve. Group effort has led to the solution of difficult problems in the past. It should be fervently applied to this all-important problem, on whose solution hinges the uninterrupted progress of civilization among mankind.

Urgent Need for Atomic Control

[1952]

Atomic bombs appear to be a lot more plentiful than was indicated by the information available a year ago. This means that atomic explosives are unfortunately much easier to make than we had any reason to believe.

Ever since the original "Baruch proposal" for international control of atomic energy introduced by the United States failed to gain acceptance by the Soviet Union, our foreign policy makers seem to have been counting on the great difficulty of manufacturing atomic explosives to give us time—time in which to improve the international atmosphere before attempting again to bring the explosive atomic situation under control.

For the moment a substantial improvement of the international atmosphere appears to be a long step beyond our current goal of attaining a position in which we can "negotiate from strength." Now it seems that at this late date it will be practically impossible from the technical point of view to design a control plan containing reasonable assurances, because the size of the atomic stockpiles cannot be accurately verified. New and ominous indications that there is a rapidly accumulating abundance of concentrated fissionable materials give fresh urgency to the problem of devising a more promising basis for prompt agreement on at least a minimum of atomic limitation or control.

There has been no official release of information on the size of the opposing stockpiles, and the only numbers available are those quoted from vague sources by columnists and other newsmen at the time tactical A-bombs were announced. These vaguely quoted numbers almost a year ago ran in the neighborhood of a thousand atomic bombs on our side and somewhat less than a hundred on the side of the U.S.S.R. Whatever the exact numbers, they are so large that there is no time to lose if we are ever to make atomic armament limitation a prime objective of our foreign policy.

The "Baruch proposal" was a very enlightened plan in a period when the United States was the only nation possessing even a few atomic bombs. It provided for quite "airtight" control of atomic production through international ownership and operation of the widely distributed production facilities. The Soviet leaders, instead, advocated essentially paper prohibitions. During the years of diatribe and propaganda there was a slight convergence of points of view when the Russians came a little nearer to discussing continuous inspection and we belatedly agreed to discuss atomic and conventional disarmament together.

New U.S. Proposal

Now there is a new UN Disarmament Commission, which is considering an "arms count" proposal submitted by the United States last April. This proposal is based on a legitimate, although perhaps unrealistic, desire to obtain complete knowledge of the world's armaments before drafting a plan to control them. It proposes a succession of stages for the disclosure of armaments, production facilities, and armed forces, followed by verification through direct inspection at each stage. The stages progress from the less secret to the most secret, covering atomic and conventional facilities at the same time.

In its present form the United States proposal is that the nations agree through a set of treaties to reveal all of their secrets and receive nothing in exchange but the other nations' secrets. Only after embarking on this procedure would they start discussing actual limitation or control. This would be a very reasonable procedure if there were mutual confidence. In this distrustful world, however, it seems necessary to find a plan which does not ask for such long-term credit, perhaps one in which it would be agreed in advance that an initial stage of disclosure would be followed by control of the facilities disclosed while the nations proceed to a successive stage of disclosure, and so on. Then each nation could know what items of control it is buying with its disclosure.

The whole question of disarmament is a very difficult one, and atomic limitation is particularly difficult because enormous power is wrapped up in such a comparatively small "package" that it is hard to detect through inspection once it has been produced by large and conspicuous installations. Each nation entering into a control agreement would have to take some risk that another nation is hiding a secret stockpile, perhaps up to some fairly definite percentage of its actual stockpile. This difficulty grows as the stockpiles grow. The most crucial aspect of the disarmament problem will be to assess correctly the relative importance of the short-range needs for improving day-to-day security by adequate armed forces and the long-range interest in avoiding the very great risk of continuing the atomic arms race until an eventual flare-up brings about a tragic setback to the progress of civilization.

The United States and the U.S.S.R. have a very real mutual interest in avoiding that flare-up. We must exploit this fact by making it clear through the strength of democracy that any plans the Russians may have for the future which would involve great risk of inciting an open conflict will not succeed, and by devising a control plan generous enough to be attractive to them and still having guarantees, perhaps not "airtight," but good enough to make the risks we would be taking considerably less than the risk of continuing the atomic arms race indefinitely. The Russians' failure to agree to our original very rigid plan does not prove that they will never agree to a less stringent one, which would still be much better than no control at all. Our policy has as yet not relaxed from insistence on the international-ownership mechanism for control of production as outlined in the Baruch proposal, although now that the stockpiles exist, the production of more atomic fuel would for a long time be unnecessary in a plan providing for a schedule of beating atomic swords into plowshares. Our policymakers thus have a long way to go in exploring the possibilities of agreement to halt the growth of the stockpiles and provide for their eventual reduction.

It is to be hoped that this difficult exploration is being pursued with vigor by the Panel of Consultants on Disarmament appointed by the State Department last spring. Theirs is an enormous job, and the panel will probably require expansion of its facilities if it is seriously to study atomic controls in addition to nonatomic disarmament. It is also to be hoped that public opinion will be receptive to the necessities of the atomic age and will not insist on traditional standards of defense which can no longer alone be adequate.

The Atom and Disarmament: Some Technical Aspects

[1953]

Kenneth E. Boulding, professor of economics at the University of Michigan, has said that the study of detail in disarmament will get us nowhere until we have a change of the temper of the times. I'm afraid as a scientist I find reasons for wanting to get down to detail, and I believe that a real appreciation of the possibilities provided by the details may change our attitude toward the temper of the times.

In taking exception with Prof. Boulding, Norman Thomas noted that, although a universal growth of peaceful sentiment would help, we haven't the time to wait for so complete a transformation before we begin to disarm. For reasons dealing with the technique of arms, I find myself in agreement with Mr. Thomas. Time *is* too short for so profound a change, so short that it is absolutely necessary to devise some settlement of world problems, agreeable and workable for both sides, now.

Most of the cold technical facts on which I base my conclusions are not new to you. Scientists and others have been emphasizing most of them for quite a while now. Not necessarily in order of importance they are:

1. The enormous destruction of the atomic bomb, especially now that it is growing up into its hydrogen version. Worse for mankind, these awful weapons of destruction can be wrapped in very small packages and are much too easily hidden. This constitutes one of the fundamental difficulties of atomic weapons control today.

2. Atomic fuels or atomic explosives (which are interchangeable) last a long time once they have been produced. That is, all of them do except tritium, which gradually disappears. The stockpile problem is with us forever, unless something is done about it. The stockpiles grow and grow and the problem grows with them.

3. When, however, we come into thinking of bringing atomic fuels and explosives under control, the third technical fact is very important— and that is that they are made in very elaborate installations. These consume a lot of power. They are easily discovered, being of a type that is fairly familiar.

4. In these large production facilities there must be some sort of clues left indicating what they have produced in the past, clues, for example,

of the general nature of radioactive tracers that are still left in the plant after it has done its job. These give us an opportunity, granting that there has been an agreement to provide access to each other's countries, to look back in the history to find out approximately what has been done, what stockpiles must be in existence.

5. When production is under way and the large plants are producing fissionable material, there is danger that some of it may be diverted surreptitiously. This was the main concern of the Baruch plan with its proposal that there should be not only inspection but international ownership of the widely distributed production plants.

6. The convertibility of fuel into explosive and explosive into fuel, or peacetime uses into wartime uses and vice versa, works both ways, and it brings us an urgent problem. If we ever get so far as to have only peacetime uses of atomic energy allowed, it leaves us the problem of making sure that there is no surreptitious preparation for effective war.

But on the credit side, it means also that in the process of transition from large preparation for war to peacetime applications only, for a long period of time it will not be necessary to make more fissionable material. We can instead take advantage of the opportunity by beating atomic swords into plowshares, of converting fissionable material in bombs into fissionable material for industrial purposes.

These six technical facts form the basis of my discussion of disarmament possibilities. Significantly, I believe they still leave the way open for a plan which could, with sufficiently imaginative effort, be capable of making clear to both of the contending parts of the world their mutual interest in survival. The way things have been getting done in the world recently, the plan would probably have to win over our own policymakers first. It then could be used in the proper diplomatic way to try to help change the mind of the Kremlin, a departure from the old procedure of waiting for the Kremlin to make up *its* mind. Let us see how the technical limitations bear on such a plan.

Having just agreed with Norman Thomas, I must now take issue with him. He said, "I would mean by disarmament, universal liquidation of weapons of mass slaughter down to a police force."

This I consider to be impossible if nations proceed with the degree of caution which I think is inevitable and, in fact, wise in a distrustful world. I have already said that there must be technical clues to tell us how big a stockpile has been produced. There is a qualification to that: It cannot be done with complete accuracy. The point needs further study, but I think there will always be something in the way of a percentage of certainty that will enable us to get a fairly good idea of how big the stockpile is. And in all probability it would be possible to gauge whether our estimate is compatible with the declared stockpile which is officially being put under control.

But no nation will be quite sure that another nation has not hidden a relatively small undeclared stockpile. To be sure, such foul play would run the risk of detection, but as long as such tactics could be effective no nation with any reasonable degree of caution should reduce its atomic stockpile to zero without a guarantee that other nations are doing likewise. Should such caution not be exerted, the relatively few bombs that might have been secreted would then be potentially decisive, providing the possibility for a surprise attack which would be overwhelming because nobody could oppose it.

It would be unfortunate to overlook this technical difficulty. The answer is not to compete in secretiveness, but to reduce the temptation to secretiveness. This could be done by agreeing that each side will retain a fairly substantial declared and inspected stockpile, compared to which a secreted stockpile would not be decisive.

I think it is important in this distrustful world that whatever agreement we do devise is not one that *invites* foul play. Thus, I would deem it technically unrealistic to try at this time to make complete disarmament an immediate aim of our foreign policy. I think a great deal can be gained by going a long part of the way toward complete disarmament, say 80 percent or so. But not as far as we idealistically would want.

Such a partial disarmament scheme would carry incentives enough to pay its own way. Instead of our present position, with the technical difficulties growing much faster than any prospect of improvement in the political climate, we would have a situation in which the technical difficulties would grow no worse while the climate might improve. This would leave an opportunity for negotiating a new agreement for more nearly complete disarmament later on.

As for possible hidden stockpiles, any disarmament plan we may devise will be imperfect. It will be limited in how far it can trust our assessment of the chances that we might be surreptitiously attacked. Anything of a purely defensive nature that we can do to reduce our vulnerability to attack is without question a step in the direction of making it easier to devise a lasting peace. It will also tend to help convince the Kremlin that agreement is the better course.

But here there is a danger. We are not yet working hard enough. We are not sacrificing enough convenience, for instance, to disperse our cities and perfect our continental radar screen. Alone, such efforts can probably do no more than postpone the prospect of the cataclysmic detonation. Coupled with intensive efforts toward a disarmament agreement, that may be decisive. Other factors that possibly could be looked upon as helpful stopgap measures might include improving the unity and strength of what we call the "free world," and even raising worldwide living standards.

Then what sort of disarmament plan might it be? I do not intend to outline here a very definitive answer, for my point is that the exploration for all promising and technically feasible variations of a plan should be a matter of intensive and imaginative study. The plan might in its main outline, for example, run along lines that have already been suggested. And yet, spelled out in detail, it may have more effect on men's convictions concerning its feasibility and seriousness, and therefore more effect on the course of negotiations.

Whatever formula is adopted should probably be a "stepwise" plan. The steps should each combine disclosure and some acts of actual disarmament, both in the field of atomic and conventional ordnance. Each step should be balanced in such a way that neither side loses appreciably in relative military potential, that is, atomic and conventional arms, geographic secrecy and production know-how.

There would be demobilization of troops and dismantling of submarines, and some of this might be balanced against discontinuance of atomic production. And there would be many other details including eventual inspection and dismantling of bombs, all carefully balanced and spelled out in advance. These steps would then gradually reveal our much cherished—perhaps over-cherished—secrets, but do it bilaterally. The details may be tedious, perhaps, but I believe the formulation of them could help convince ourselves and others that we all have more to gain than to lose in the process.

In relation to the question of national sovereignty, a disarmament plan should be designed that would end the arms race as soon as possible. We cannot wait for the nations to make more than the minimum sacrifice of sovereignty consistent with reasonable guarantees of good faith. For one reason, the atomic aspects of the problem will become overwhelmingly difficult with further delays.

The establishment of the plan doubtless would mean lifting the Iron Curtain, or opening large holes in it. The Iron Curtain goes far beyond the conventional exercise of national sovereignty, and there are many who say that there is no hope of lifting it even a little.

This is said on the basis of the past seven years of frustrating attempts at negotiation. The judgment is in many cases influenced by Soviet manners, by the maddening way in which the Soviet front man is taught how to say ''no.'' Former Ambassador Benjamin Cohen has told us something of this. As U.S. representative on the UN Disarmament Commission, he endured Soviet pressure long and patiently in a forum where more genuine attempts at negotiation should have been possible had the Soviet leaders already made up their minds that they had an interest in disarmament.

But the picture is not entirely one-sided. If it was evident that the Soviets were not particularly interested, the United States correspondingly did not go very far in making proposals, nor did it seem eager to do so. With better advance preparation of detailed proposals, I believe we would have done better to go a little further.

My complaint with the Truman administration was that, having made one good try, it stopped. It acted as though it had tried all possibilities and then proceeded to waste the best years of our lives, content to hold the line for the present without further concerted effort to devise a plan that might halt the advancing atomic threat before it is too late.

There are no clear indications of the attitude to be expected of the new administration. Secretary of State John Foster Dulles in some remarks last October suggested that he considered it too late for negotiation, but he has sometimes seemed more optimistic.

General Eisenhower has been known as a negotiator in the past. It may be remembered that when he was chief of staff some years ago there was a committee of three colonels whose sole assignment was to think about the problems of long-range security. If that sort of thing was in his mind then, I can't help thinking that the type of intensive technical-political study needed to prepare the way for long-range security through disarmament might appeal to him now.

There is too much talk about getting into a good negotiating position, this without having developed the technical basis for knowing what we want to negotiate when we get there. There remains an urgent need for mounting a real effort to study and to formulate all aspects of the armament control problem, and to synthesize the possibilities into the best workable control plan to the mutual advantage of both sides. Since the Soviets won't work with us on this sort of effort in the UN—and that's about all we've really learned from their vehement attitude there—let's design the product, then work in on the sales end.

The Stassen Appointment: Turning Point in Disarmament Thinking?

[1955]

The recent creation of a cabinet-level post, filled by Harold E. Stassen as President Eisenhower's Special Assistant on Disarmament, is a dramatic event. It signalizes a turn in our government's show of interest in thinking through the enormous problem of finding a sane alternative to the risks of the unlimited arms race.

To be appreciated, the event should be viewed in the appropriate historical perspective. The year 1945 saw the birth of the United Nations at San Francisco, innocent of the atomic facts of life disclosed later the same year at Alamogordo, Hiroshima, and Nagasaki. Man was suddenly faced with the problem of surviving the means of his own destruction, in being or in foreseeable prospect. Yet without awaiting a solution, America demobilized the world's greatest fighting force.

Nineteen forty-six was the year of genuine attempts at a world political adjustment. In an inspired search for ideas, the five prominent and able men of the Acheson-Lilienthal subcommittee of the State Department worked hard together for six weeks and suggested an international managerial type of control of atomic facilities, to avoid having any dangerous atomic materials at large. This was the first and apparently the last time such concerted thought was put on the atomic disarmament problem in official circles. As a consequence, the Baruch proposal was made in the UN, a most remarkable offer to give up our monopoly under appropriate conditions, its dramatic edge only slightly dulled by Mr. Baruch's conservative overemphasis of the veto question. The protracted Soviet refusal to enter into the spirit of genuine disarmament discussion made the future look dark.

As a reaction, the year 1947 was the downward turning point of American thought on disarmament. This we learn most definitely from the testimony before the Gray Board by Osborn,[1] speaking of the time in March 1947 when he took over Baruch's assignment as U.S. representative on the UN Disarmament Commission. After outlining Oppenheimer's warning to him that there were dangers in continuing negotiations, both because we might make undue concessions without getting rid of the Iron Curtain and because of the propaganda value to the Russians, Osborn said:

> I went back to New York and I saw McNaughton, the Canadian representative . . . and Cadogan, the British representative, and Parodi,

the head of the French delegation. . . . They all felt very strongly that the negotiations should continue. They said they really had not [had] a good look at the Baruch plan, they had not taken much part in drawing it, they did not know what it would look like if it were put in more detailed form. They said they would be in an impossible position in their own countries if they agreed to call off the negotiations. . . .

I asked [Senator] Austin if I could go to this meeting [of the President's Executive Committee on the Regulation of Armaments] with him in Washington. He said yes, he would take me along. Austin felt very strongly that we should continue negotiations (and said so at the meeting). Forrestal said, "This is a lot of bunk," and so did Patterson. . . . He said we should not go on with the negotiations.

. . . I said I agreed with Austin that we should continue the negotiations for quite different reasons. I felt the Russians had no intention seriously and they would not agree to any form of control that we could accept, but that I had talked to . . . the British, the French, and Canadian representatives and these men were very insistent that we continue negotiations.

I thought if we were properly on our guard we need not make any bad mistakes or endanger the situation, and it would be very injurious to our international position to take a lone position, refusing to negotiate.

Forrestal said, "That makes sense to me: what do you think, Bob?" Patterson said, "I think we should go ahead if this is the reason and if we do it with our eyes open." Acheson said he was opposed to our going ahead. Lilienthal said that he agreed [with going ahead]. Acheson said, "If you feel this way, it is all right . . . to go ahead."

Oppenheimer's warning might be interpreted as advice to stop talking and start thinking again. But the decision was thus taken to go on talking. The talk continued from 1947 to 1955, with much patience in enduring frustration, with a little convergence of expressed points of view, but without evidence of further thought of a depth commensurate with the importance of the disarmament problem.

During all this time there have been many eruptions in the world's local tensions to distract all statesmen from thoughts of long-term settlements. In Europe, where the line between East and West has been held since Czechoslovakia, there have been the Berlin airlift and the impact of the Marshall Plan and NATO. In Asia, where communism has made great gains, the series of crises has not yet forced us into the use of atomic weapons, but the realization that it might at any time do so highlights the question of an overall settlement.

The advent of the H-bomb in both East and West in 1952–53, with its new order of magnitude of destructive danger, has brought a new sense of urgency in the expressions of concern by the world's leaders, especially by President Eisenhower. Yet it has brought no apparent intensification of studies of disarmament possibilities. Existing officials and agencies have been too preoccupied with immediate problems, and no new agency free to devote itself to the question was created until the recent Stassen appointment.

During all this discouraging time it has seemed to many people that with the other side unwilling to negotiate there was no use wasting time with serious thoughts about disarmament, and to others it has seemed that at least one realistic disarmament plan, if it could not be arrived at through negotiation, should be formulated in reliable detail by careful and imaginative study on our side. This would provide a basis on which our leaders could reasonably decide whether disarmament really makes sense.

It is like the chick-and-egg question. If the will to negotiate does not come before the formulation of a plan, then we should formulate disarmament plans expecting that negotiation may develop from the improved understanding of disarmament possibilities. If we could find a sufficiently attractive proposal, our foreign policy could be oriented accordingly, in the hope of "selling" it to the other side on its merits.

Up until now, the former view has prevailed in our government, that the will to negotiate must come first. The Stassen appointment suggests a shift to the latter view, or at least that Mr. Stassen has been put in a favorable position to promote the latter view if he lives up to his great opportunity.

To do justice to the challenge, on which may hinge our future survival, will require the creation of a very special sort of staff, rather different from most regular government agencies. It has frequently been suggested, at earlier stages when the problem was easier but its importance less widely appreciated, that there should be a high-level commission charged with disarmament planning. If properly handled, Stassen's cabinet-level post should have administrative advantages over a commission and at the same time all its advantages, the broad scope and the high productivity of capable minds stimulating one another. It is to be hoped that Mr. Stassen will have the breadth of vision and the persuasiveness to attract some of the ablest men from the reservoir of the entire nation, not just from official Washington. Experts who already are acquainted with certain aspects of the problem will be needed, and some of these will probably be most easily available from established government agencies, but there would be danger in loading the staff too heavily with these because of attitudes developed during the long period of discouragement.[2]

Mr. Stassen has recently complained of the "congealed cynicism" with which he finds himself surrounded. The complaint itself is indicative of his healthy attitude and long interest in the problem of finding a world settlement. May he collect the right kind of men around him and build a fire under them!

The job to be tackled is twofold, hardware and politics. A very complete evaluation of the technical possibilities will have to be made as a basis for deciding what military hardware we can afford to do without under various conceivable world politico-economic conditions. Likewise, a complete understanding of political and sociologic conditions will have to be used to decide what politico-economic incentives and assurances may be brought to bear to convince ourselves and the rest of the world that we will all be safer without the hardware. The staff must be constituted accordingly, giving opportunity for the flexible interplay of the demands of these two factors in the search for the most favorable overall proposals. This will be the first time (since 1946, at least) there has been a place in our government for the interplay of these two kinds of ideas. It will be the first time there has been in high places anyone with the mandate and the undistracted time to devote to disarmament thinking. The time is very late. The job must be undertaken without the assurance that acceptable answers exist, but with the determination that they shall be found if they do exist. Armchair opinions that none do exist must not be allowed to stand in the way of a determined and all-out search. The scale and the value of research and development projects, often carried out by private contractors for government agencies, must be borne in mind when setting up the staff. Perhaps outside help with parts of the investigation should be arranged, with some of the "second laboratory" type of overlap providing the spur of competition.

In a letter to the editor in the *New York Times* of February 6, some time before the Stassen appointment, Szilard[3] eloquently underlined the urgency of finding a basis for agreement to avoid the danger, which he considers immediate, and remarked "To outline such an agreement in some detail will require the kind of imagination and resourcefulness that cannot be expected from the government. In our political system the intellectual leadership needed here can arise only through private initiative." This thought must constitute a constructive challenge to Mr. Stassen. As a sound criticism of the work of routine government agencies, it must emphasize the necessity that he create a very unusual and inspired organization.

At the same time the Stassen appointment lends purpose to the exercise of private initiative to the same end, with the conviction that an independent effort can do better, or can at least increase the thoroughness of the exploration of all possible angles. Before the appointment, there would have been doubt that a good idea from a private effort would find a sympathetic hearing within the government. With the Stassen position as an expression of earnestness and as a natural channel, this doubt is greatly reduced. Now is the time for good men to develop and contribute good ideas. It might now be possible for organized private effort to arrange the type of sponsorship that would give it special access to the secret information needed in this field in which sound assessment of modern weaponry is essential.

Our present foreign policy stands committed to whatever security there is in the threat of massive retaliation by superior nuclear force, merging gradually into the fragile mutual deterrence of opposing and equally massive threats. The eventual fatal slip seems almost inevitable. It must be the first purpose of the Stassen effort to provide a sound alternative capable of convincing our national leaders that it is in our national interest to take down our guard as others do likewise.

The Stassen group must be given time to work with calm intensity, free from political pressures or the demands of publicity. So long as the quality and size of the appointed staff indicate that an adequate and all-out attempt is being made, no further report or evidence of progress must be expected for quite a period of patient waiting.

Independent efforts to improve the world climate and to prepare public opinion for the acceptance of unprecedented international arrangements will be of great help. Informed and insistent pressure for disarmament from the thinking portion of the public can help assure a sympathetic reception of the eventual Stassen proposals in appropriate parts of the government. However, final preparation of the broad base of public opinion, in which the Stassen organization might be expected to take some part, will still be needed after the decision in high governmental quarters that an advantageous solution of the problem exists.

If a really valiant try should be made in vain, no stigma should be attached to failure, for we cannot be sure in advance that a solution does exist (just as we cannot be sure that it does not). The problem must not be allowed to go by default. It seems that human intelligence must be able to find a way out of the mess into which its own technical ingenuity has brought the human race, and the Stassen group should be the thinking organ to do it.

NOTES

1. *In the Matter of J. Robert Oppenheimer* (Washington: U.S. Government Printing Office, 1954), p. 344.

2. According to a *New York Times* dispatch from Washington dated May 14, "Harold E. Stassen, special assistant to the President on disarmament matters, has completed his staff of advisers. He disclosed the eight members of his staff today. They are: Edmund A. Gullion, State Department; Col. R. B. Firehock, Army; Capt. D. W. Gladney, Navy; McKay Donkin, Atomic Energy Commission; Col. Ben G. Willis, Air Force; Lawrence D. Weiler, State Department; Robert E. Matteson, and John Lippman, both of the Foreign Operations Administration." It thus appears that Mr. Stassen has, at least initially, chosen not to call upon talents from the country at large.

3. Dr. Szilard's letter appeared in the *Bulletin of the Atomic Scientists* 11 (March 1955): 104.

Arms Control Effort Buried in State

[1957]

With the demotion of the "Special Assistant to the President for Disarmament" from independent cabinet status to an innocuous position in the Department of State, we have further evidence of the government's myopic smugness in face of a precarious future. It is one thing for the government to say that there is no easily acceptable alternative to its present policy, and another one to announce, in effect, its intention not to seek seriously an alternative to a perpetual nuclear arms race among many nations.

For a dozen years the growth of nuclear weapons has been outstripping by far the development of political arrangements to avoid disastrous war. The deterrence they provide seems essential to short-term stability, but dangerously precarious in the long run. The threat is to the continuation of civilization itself. Weapons become faster and faster and thus more difficult to bring under ultimate control. The danger of accidental flare-up will increase as surprise becomes more devastating and as more nations get into the nuclear race. The devising of effective and mutually acceptable worldwide limitations is an enormously difficult job, requiring much more searching effort than has yet been expended. The organization of a sufficiently resourceful and devoted attack on this problem would require insistent initiative at a high level, probably by the president himself.

There seemed hope that the president had recognized the need and made a beginning toward adequate organization when, two years ago, he announced the creation of the office of Special Assistant to the President for Disarmament, and appointed Governor Harold E. Stassen to fill it. Skeptics could point to the obligation to find Stassen a job, and to the propaganda value of the much-touted appointment, but the president's sincerity was attested to by the fact that Stassen was outstanding among the prominent men of the administration for his deep and long-continued interest in the possibility of attaining greater security through mutually advantageous international accommodation. The president was doubtless beset by conflicting advice concerning the desirability or feasibility of arms limitation of any sort. It seems likely that he really wanted progress in this field, if a scheme could be quickly found on which his administration team could agree, and that he considered the Stassen appointment the right way to find out. If so, he appears to have had little idea of the magnitude of the effort required or of the need for positive leadership in this exploratory and unconventional search, and the recent transfer of the

Stassen office to the State Department seems to indicate that he has already taken "no" for an answer.

Through all the years of discouraging disarmament negotiations, it has been a matter of guesswork to what extent the various national proposals may have represented sincere aspirations of governments and to what extent they have been intended merely to give the impression of purity of heart. By and large, the latter intention has seemed to prevail. In our government, even at the most favorable times, those in high places could not agree on a policy envisioning renunciation of military strength in recognition of similar renunciation by others. The last two years, when we have had the Stassen office, have probably seen the most nearly favorable consideration—or at least the most extensive consideration—of arms limitation possibilities within our government since 1946. The one possible exception, at a time when the technical possibilities were more favorable, was the briefly considered and quickly dismissed Bush proposal to seek agreement to desist from H-bomb tests just before our first one was carried out in 1952.

The record of the past two years is not one of bold initiative or marked accomplishment, but of a somewhat intensified exploration of arms limitation and international accommodation possibilities. It seems likely that the initiative for the summit conference at Geneva came from Stassen at about the time his office was being created, and if so, this, and the accompanying attempt to improve the climate for negotiation, was probably his boldest gambit. That it could not be implemented by proposals less timid than the early warning system in the "open skies" plan was due to the failure of the administration team to agree on the desirability of restricting the headlong arms race, even when urged by Stassen. The most glaring failure of the administration was its failure to accept, with no strings attached—or, at least, to explore—Bulganin's proposal to establish ground check posts for observers at key points throughout large countries. This alone would have provided a rudimentary early warning system, and if used as a check on weapons tests, could have served as the beginning of very real limitations on arms development. This then is the measure of our lack of policy—we don't even try to take the surprise out of deterrence.

The present announced U.S. policy in the arms limitation negotiations, even though it shows only a small advance over our earlier policy, does show that the Stassen office had made some progress before its demotion. This progress may even have hastened the demotion! The first two points of the policy appear to be window dressing. They are: (1) that we would like to arrange for diverting all future atomic production to peaceful uses, if adequate controls and inspection can be established and (2) that if this is done we will consider cessation of H-bomb tests. This is a graceful concession to world opinion opposing tests. On casual reading, it seems to say that we favor stopping tests, if adequate inspection can be set up. It really says that we will agree to stop tests only if extremely detailed inspection can be set up, adequate for the more remote goal of stopping weapons production—an inspection which can be dismissed by the ruling "realists" as impracticable for the present. It is the administration's way of saying "no" to further consideration of the sensible proposal to stop tests with adequate monitoring. We are not even planning to explore the sincerity of the Russians' offer to stop tests. These first two parts of our present policy show what little progress the Stassen office has made, how two more years have been nearly wasted on the political side of the problem while rapid advances have been made on the technical side—to the point where we can now assure Britain of delivery, within "much less than five years," of large numbers of fifteen-hundred-mile missiles carrying hydrogen bombs.

The remaining three points of the present policy appear to be at least intended for serious negotiation, and represent slight progress. One point concerns a reduction of the number of men in the armed forces, although a smaller reduction than is favored by the Russians, who would like to see a manpower shortage make us give up our overseas bases. This point is relevant to the atomic arms problem mainly as an indication of our growing dependence on the myth—or, at best, the hope—that small wars can be fought and won by the limited havoc wrought by "tactical" atomic weapons, without growing into big wars. Another point calls for establishing a system of registering nuclear weapons tests in advance and permitting limited observation by other nations. This step in itself would do nothing to control the rate of development of future weapons; but as a first step toward the establishment of an international climate in which such control might become possible, it is well worth taking and should be a minimum accomplishment of the present round of negotiations. But there is also a further point, phrased so vaguely as to permit enormous scope for negotiation if favorably interpreted: international control of "outer space" missiles. This could even mean a frontal attack on the problem of preventing the development of intercontinental ballistic missiles (ICBMs) and even intermediate range ballistic missles (IRBMs) before it is too late. The vague wording might indicate that administration policy is still fluid on this question. Even without the recent transfer of the Stassen office, actual negotiations of course would have been carried out under the general direction of the Department of State; but the transfer removes Stassen's freedom to try to bend administration policy in this direction. In any case, the timid wording of the other points does not encourage the hope that such a bold construction can be put on this one. "Outer space" missiles will probably be interpreted literally—as artificial satellites; but even this limitation of independent national technical development, in a field possibly related to future armaments, would represent progress toward reducing the danger of an independent breakthrough, in keeping with the spirit of Senator Anderson's April 1956 speech.[1]

The White House announcement of March 1 (*New York Times,* March 2, page 1), transferring the Stassen office to the Department of State, has received scant notice and little comment, yet with it has passed into oblivion the one small effort which, with sufficient insistence from the president, might have grown into an adequate pursuit of the hope to avoid, through rational and mutual arms limitation, an endless—or eventually disastrous—arms race.

The appointment now appears to have been merely a trial balloon. It showed that it is not easy to unite the viewpoints of the various administrative departments on this unprecedented problem. It was a very personal appointment. It happened when Mr. Stassen was conveniently available, and he is being pushed out of it with respectful decorum—his political amibitions are growing in other directions.

Perhaps the appointment never implied the recognition of the deep national need for long-range investigation of the possible foundations of a future armament policy, independent of the exigencies of the daily execution of foreign policy. It was much too intimately associated with negotiation to serve adequately in this exploratory capacity. It was never adequately staffed for independent investigation of the various facets of the problem. It did not possess the necessary technical capabilities, and, after framing the questions relevant to a particular approach to disarmament, it had to rely on answers from its eight part-time advisory "task groups." It would have had to grow to become adequate. Instead, it has virtually vanished, save as an instrument of negotiation. If an adequate investigating and policy-searching project in the disarmament field is ever to arise, it must again start from scratch. In starting later, it will face even tougher problems.

NOTE

1. See *Bulletin of the Atomic Scientists,* June 1956, pp. 223–26.

2 Test Ban

Before the advent of the H-bomb in 1953, tests of A-bombs in Nevada and the South Pacific were causing troublesome radioactive fallout, against which there were widespread protests. There was public clamor for a test ban to stop fallout and, in the United Nations, the delegate from India proposed cessation of such tests to avoid fallout and reduce tensions while disarmament deliberations were in progress. It thereafter came to be appreciated that an agreed-upon ban on the testing of nuclear explosives, in addition to ending the nuisance of fallout, could serve the even more important purpose of greatly slowing down the development of more sophisticated nuclear weapons and thus serve as an effective means of arms control through arms development control.

There seems to have been no discussion of this arms control aspect of a test ban in the public print until the suggestion appeared in the first two of the following papers in mid-1954. At that time deterrence depended on delivery of A-bombs or the relatively new and primitive H-bombs by manned aircraft against which some defense was possible. The emphasis in those papers was on impeding or preventing the development of more sophisticated H-bombs for swift delivery by intercontinental ballistic missiles, an offensive threat under which we now live and against which there is no effective military defense.

It was not then generally known that a similar proposal had been made at a more propitious time two years earlier behind the veil of government secrecy. That such a profoundly sensible and uniquely opportune proposal at a high level should have failed of acceptance is a measure of the narrowness of purpose in our decision-making process, bent on keeping ahead of the Soviets in the short term regardless of long-term consequences. The suggestion had been made directly to Secretary of State Acheson by Dr. Vannevar Bush, former director of the wartime Office of Research and Development and chairman of important committees thereafter. Word of his initiative was made public during the course of the sordid hearings that resulted in stripping Robert Oppenheimer of his security clearance, McCarthy-era hearings in early 1954, the transcript of which was published later that year. One of the charges against Oppenheimer was that he had, in concert with several other prominent scientists, advised against going ahead in 1949 to develop and test a thermonuclear explosion in an awkward device that was indirectly a precursor to an H-bomb but was far from being an H-bomb. In this connection Dr. Bush, testifying in Oppenheimer's behalf, was asked to explain why he himself had advised against carrying out the first actual H-bomb test in late 1952.

> There were two primary reasons why I took action at that time, and went directly to the Secretary of State. There was scheduled a test which was evidently going to occur early in November. I felt that it was utterly improper—and I still think so—for that test to be put off just before election, to confront an incoming President with an accomplished test for which he would carry the full responsibility thereafter. For that test marked our entry into a very disagreeable type of world.
>
> In the second place, I felt strongly that the test ended the possibility of the only type of agreement that I thought was possible with Russia at that time, namely, an agreement to make no more tests. For that kind of an agreement would have been self-policing in the sense that if it was violated, the violation would be immediately known. I still think that we made a grave error in conducting that test at that time, and not attempting to make that type of simple agreement with Russia. I think history will show that that was a turning point when we entered into the grim world that we are entering right now, that those who pushed that thing through to a conclusion without making that attempt now have a great deal to answer for.
>
> That is what moved me, sir. I was very moved at the time.[1]

This was a unique opportunity lost. Then the incentive to agree not to test was great, for the potential reward was great, avoiding entry into the H-bomb age, an all-or-none decision. When later the idea was encountered that powerful tests could be carried out underground and further that the shock wave could be muffled by putting them in very big holes underground, special agreements would have been required to monitor the lack of testing, but the incentive to agree on them would have been correspondingly great. After this opportunity was missed and testing proceeded, H-bombs became increasingly sophisticated and the incentive diminished to stop testing to prevent further improvement. Yet it made sense, and still does, to make a comprehensive test ban agreement for arms control purposes. Modern technology has by now advanced to such a point that it is also possible to monitor from afar long-range missile trajectories, so a ban on missile tests could be usefully combined with a ban on tests of nuclear explosives.

The quest for a comprehensive test ban since then has been long and arduous. A test ban was recognized as the easiest minimal step toward arms control. The long hassle over it served to deflect public attention and in a sense excuse the government from more serious and direct pursuit of arms control. Adlai Stevenson in the 1954 presidential campaign proposed not only a partial test ban treaty but also a de facto moratorium on H-bomb testing. Such a moratorium was later agreed upon with the Soviets to be in effect for one year while the 1958 test ban negotiations were in progress. At the end of the year President Eisenhower formally ended the moratorium by declaring that we would no longer be bound by it. The Soviets then prepared for testing while we did not do so on a standby basis as a precaution. We were taken by surprise and were unprepared to match them when after a year and a half of calm without a declared moratorium they carried out a series of powerful tests. This act was widely heralded in the West by proponents of a continued arms race as a perfidious breaking of the moratorium and a demonstration of the futility of making treaties with the Russians, despite there having been no treaty or even an informal moratorium agreement to be broken.

The problem of seismic detection of underground tests was the foremost concern by 1958 when that conference of American and Soviet experts at Geneva made reasonable recommendations for dealing with this. Their proposal may have seemed likely to be adopted, perhaps dangerously so to proponents of further testing who torpedoed it by inventing big-hole muffling. In the early sixties our negotiating position was for a ban on tests in the atmosphere to avoid fallout while the Soviets held out for a comprehensive test ban having arms control implications, until in 1963 they conceded. Despite the technical improvements and some political efforts in the meantime, we still have only the partial test ban of 1963. Since that time, there have been many more tests underground than there had been altogether up to then. Achievement of the partial test ban may be seen as a result of public pressure that centered on the fallout issue. Public pressure can have such an effect, but in this case it was effective only because the Pentagon tolerated a compromise that preserved the possibility of continued testing underground.

To some of us concerned about arms control, achieving the partial test ban seemed unfortunate. Fallout being the sort of close-to-home issue that arouses public concern, it had been hoped that public pressure based on fallout could lead to a comprehensive test ban effective also for arms control.

Further efforts during the sixties to achieve a comprehensive test ban were stymied by the possibility of big-hole muffling of underground tests, but by the seventies, through concerted government-sponsored effort on this problem, the techniques of seismology had improved sufficiently to make possible a test ban adequately monitored down to explosive yields so small as to be relatively unimportant. Since then negotiations have failed to achieve such a ban, not through technical difficulties but through lack of political will.

Top public officials quite properly feel responsibility for maintaining reasonable military strength, but some who come to office with a zeal for such measures of restraint as a test ban are too easily impressed by imposing briefings by the military branch, and the needed political will is lost. At least three presidents, due to personal conviction and the urging of scientific advisors, have wanted to achieve a ban on all nuclear tests—President Eisenhower urged by science advisor George Kistiokowsky, President Kennedy by AEC Chairman Glen Seaborg, and President Carter largely on his own initiative—but all have been dissuaded. Two of them were dissuaded by visits of scientists representing the two main arms development laboratories presented to the president by high administrators, President Eisenhower by Edward Teller and Ernest Lawrence, associated with Livermore and presented by Lewis Strauss, as mentioned in chapters 1 and 7, and President Carter by Harold Agnew and Roger Batzel, directors of Los Alamos and Livermore Labs, presented by Secretary of Defense James Schlesinger. There have, of course, been many other influences both ways but the similarity of these two instances many years apart is a striking reminder of how much things remain the same over the years.

The arguments against a test ban that seem to impress presidents and members of Congress alike are always based on what we need in our weapons program: we need to develop that most recent idea for a new weapons system, we need to redesign and modernize our oldest weapons and of course test the modernized designs, we need to proof test samples from our stockpile occasionally to give us confidence that they have not deteriorated, for we might as well not have the weapons if we cannot rely on them to function properly. The emphasis is on what *we* need for modernization and reliability of *our* weapons with too little regard for what happens on the other side. The idea of a test ban is of course to limit both sides approximately equally, to interfere as much with what *they* can do to achieve the modernization and reliability that *they* desire. If inability to test does indeed lead to uncertainty about performance as well as slower development, this attests to the effectiveness of a test ban in stabilizing the balance. Military judgments tend to be made by a worst-case analysis. When there is uncertainty of performance, a prospective attacker should be effectively deterred by considering the worst case that the weapons of the other side might work more reliably than his.

The six articles in this chapter indicate something of the mood of the times from 1954, when the idea of a test ban to slow down the arms race was new, to 1963, when the partial test ban treaty was signed. We see acceptance of the basic assumption that we must be very much on guard against Soviet attack and emergence of the feeling that there were already enough nuclear weapons for deterrence, particularly with our vast superiority. We see how President Eisenhower, having been dissuaded, had to debate against a test ban with Stevenson, who introduced it as a campaign issue. One notes how difficult it was to get even as mild a measure as the partial test ban through the Senate, and that I underestimated the effectiveness of underground tests for further development. With the size of stockpiles starting on an exponential rise, there was a sense of urgency to make progress with arms control to stop it before it would get out of hand—as it is now. We see the reluctance to forgo arms development then, putting us in greater danger now. In that perspective we should see our reluctance now.

NOTE

1. *In the Matter of J. Robert Oppenheimer*, transcript of hearing before the Personnel Security Board, AEC, Washington, D.C., April 12 through May 12, 1954, p. 562.

H-Bomb Control: Safeguarding the World

[1954]

Before the UN Assembly last December, President Eisenhower spoke eloquently of our desire to seek "more than a mere reduction or elimination of atomic materials available for military purposes" and made his famous atompool proposal for adapting these materials to the "arts of peace" on an international basis. The proposal met Soviet rejection at the time and failed to reduce East-West tension, but its failure does not mean that a carefully devised, practical, and mutually advantageous arms limitation proposal, if made now, would meet a similar fate.

Addressing the Supreme Soviet early this year, Malenkov expressed the realization that both sides, not just the "capitalist" powers, stood to lose an atomic war. This was the first indication that the magnitude of the threat of mutual annihilation had penetrated the consciousness of the Soviet leaders. Preliminary reports of the recent secret meeting in London of a UN subcommittee on disarmament unfortunately suggest that no unusual effort was made to exploit the new awareness, and these meetings seem to have followed the frustrating pattern of earlier, more widely publicized negotiations.

The last of those earlier negotiations took place in the UN Disarmament Commission in the spring and summer of 1952. By then the West's proposals had changed only a little from their original 1946 form, the "Baruch plan." They were still based on that plan, although it was already obsolete in that it paid too much attention to avoiding the new production of atomic weapons and too little to accounting for past production. The new feature proposed by the United States in 1952 was an "arms count," a plan for the progressive (in five stages) disclosure and verification of the extent and nature of atomic and conventional armaments by both sides. This was based on the idea that one cannot talk about armament reduction without knowing what armaments exist, but it was unrealistic because it demanded great sacrifices without providing any incentive except the hope of further agreement.

The Soviets' position had also changed but little from their earliest proposals. They still demanded starting off with a blanket prohibition of atomic weapons, though they had relaxed on the question of atomic control and admitted that inspection, in some ill-defined way, must be continuous. They opposed the idea of proceeding by stages in order to build up mutual confidence. Each side seemed to consider its proposals perfectly safe in the sense that they stood no chance of being accepted by the other side or of stopping the arms race, which was politically expedient and did not yet threaten catastrophe.

This was the situation when the French delegate, Jules Moch, made his valiant attempt to reconcile the opposing sides with a proposal similar to that which he outlined recently in these pages ("Banning the H-Bomb—A Feasible Program," by Jules Moch, *The Nation*, May 15, 1954). Instead of five stages as proposed by one side or a single-step introduction of prohibition and controls as proposed by the other, Moch advocated a three-stage plan in which the "arms count" feature was dovetailed with steps of actual disarmament.

As he presented it in *The Nation*, the French scheme suggested, along with limitation of conventional armaments, that atomic production be stopped while international inspectors got acquainted with the production plants. In the last stage production would be resumed and all the nations would trustfully beat atomic swords into plowshares. This seems like the right spirit in which to go about disarming, and a good logical order in which to do it among friends, but it also shows how superficially the statesmen most concerned have considered the problem of designing a mutually advantageous disarmament plan capable of functioning in a distrustful world.

Besides outlining the French proposal, M. Moch in his article made the personal suggestion that it might be relatively easy to get an agreement to ban H-bombs and control A-bombs. But he based this view on two serious misconceptions. "It is not unlikely," he said, "that before long nobody will dare to continue experiments with the H-bomb." In reality, neither of the principal contenders in the armament race will dare not to so long as no agreement is reached. He then suggested that it would be practicable to "prohibit the use" of H-bombs because the detonation of an H-bomb can be detected thousands of miles away. In reality, although the ease of distant detection does make practicable a ban on the *testing* of H-bombs, it provides no check at all on whether they are being stockpiled.

M. Moch was at an obvious disadvantage, as is any statesman not representing one of the two chief antagonists, in not having access to all the atomic facts of life on which a practicable proposal must be based. Still, it is a pity that some statesman not directly involved in the atomic armaments race has not pushed more detailed proposals, since the representatives of the "atomic colossi," to use Eisenhower's words, have apparently been too preoccupied with "malevolently eyeing each other." In particular, American and other Western statesmen are preoccupied with preventing further expansion of the Soviet orbit as a necessary precondition to confronting the Kremlin with a choice between reasonable agreement and mutual atomic annihilation. They do not seem to be thinking much about possible types of agreement or to be trying to hasten an overall settlement by coupling it with an attractive disarmament proposal including atomic weapons.

Since the technical problem of designing a possible agreement with adequate guarantees is becoming rapidly more difficult with the accumulation of atomic stockpiles, it is dangerous at this crucial time to have a virtual moratorium on creative thought about possible agreements. Our country should have a high-level group of imaginative thinkers, perhaps a successor to the State Department Panel on Disarmament of 1952, at work on the problem of devising and evaluating possible reasonable agreements in all details. Our diplomats have waited too long for the other side to change its mind spontaneously and agree about the general approach before worrying about the details. It is high time to see if a complete agreement cannot be devised so mutually advantageous as to make the Soviet leaders change their minds. An adequate plan may not be still possible, but the stakes are high enough to make the gamble worthwhile. Since the United States seems to be letting the problem go by default, other countries might well take a try at it. They might do better than we could in devising something acceptable to both sides. At the very least it would be interesting to have M. Moch fill in some of the details of the French proposal and see how it shapes up.

It seems likely that if the arms race can be stopped at all, it will be by an agreement on some imperfect degree of arms limitation, a vigilant truce in the arms contest, so to speak, rather than a plan for complete and universal disarmament. More would be preferable, but this would be better than continuing the unlimited arms race. It would end the dizzy spiral and the continual temptation to be the first to strike. It would keep the number of atomic powers from growing indefinitely. It would prevent the technical difficulties of control from getting continually worse and thus preserve the possibility of a more far-reaching agreement.

There remains for consideration one much less ambitious arms limitation proposal that would be fair and advantageous to both sides because it would gradually reduce the dangerous preponderance of offense over defense. This is the proposal to ban further *testing* of H-bombs by agreement among the powers, and to set up an international monitoring agency. Access to the territories of the countries principally concerned would not be necessary in view of the far-reaching meteorological effect of H-bomb tests. The test ban was suggested by India, without mention of monitoring arrangements, in the hope that it might serve as a "standstill agreement" pending further disarmament, but it would make an important contribution to world stability even if no further agreement should follow. It would seriously hamper the development of more easily delivered weapons of offensive power while permitting the development of defensive power to go on unchecked. And it would do this equally for both sides. If the statesmen of the world are too discouraged to initiate a serious search for an agreement on even partial disarmament, they should at least pay some attention to this positive possibility of slowing down the approaching doom.

Ban H-Bomb Tests and Favor the Defense

[1954]

President Eisenhower said to the UN Assembly last December that to do nothing but emphasize our armed might "would be to accept helplessly the probability of civilization destroyed . . . and the condemnation of mankind to begin all over again the age-old struggle upward from savagery toward decency, justice, and right." Malenkov before the Supreme Soviet a little later expressed the realization that both sides, not just the "capitalist" side, stand to lose an atomic war. From this it appears that the heads of state recognize the danger of going on with the atomic armament race; yet there has been no progress toward arms limitation agreement. The recent secret disarmament talks in London have sadly followed the pattern of their goldfish-bowl predecessors. The Soviet representatives did not budge perceptibly from their earlier position.[1] The Western representatives did abandon the Baruch plan provision for international ownership that has long been obsolete because of the stockpiles. They tried, but only halfheartedly, for their try was not based on the sort of exhaustive formulation of the control problem and the balance of incentives that might have *some* chance of inducing a change of Soviet policy.

This failure, though ominous from a long-range point of view, is perhaps not surprising. Viewed from the usual level of myopic practical politics there is something incongruous about expecting that two nations now as distrustful of each other as the United States and the USSR should agree to a complicated apparatus of disarmament and atomic control, with inspectors performing carefully prescribed tasks and interfering with some activities throughout both countries. Yet any significant degree of disarmament cannot reasonably be based on less. No responsible government is going to relinquish its critical armaments, upon which it depends to discourage aggression, unless it has very real assurance that other states are doing likewise. Paper prohibitions do not provide real assurance.

Opportunity for Arms Limitation

There remains one opportunity for a valuable degree of arms limitation by mutual agreement that does not require any such complicated and seemingly incongruous apparatus of internal control. It is an agreement to prohibit further tests of H-bombs and to have the long-range monitoring done by an international agency so as to guarantee that any violation would be unequivocally announced to the world.

Even though it provides no disarmament and provides arms limitation only indirectly by limiting the development of new types of arms, this agreement would have a very important value to both sides. Its chief value is that *it would slow down the rate of development of new techniques of offense and allow the techniques of defense to come closer to catching up.* The most explosive feature of the dangerous international situation arises from the undisputed supremacy of offense over defense. To lessen this without giving either of the contenders a distinct advantage over the other is to make a more stable world.

An objection may arise that H-bombs are already so horrible that no amount of testing will make them much worse, for they can already destroy the largest cities. As a result of the first six or eight years of testing of A-bombs, we have heard of vast improvements mainly in the versatility and ease of delivery, partly as a result of improved efficiency. Presumably most of these improvements could not have been achieved without experimental tests. It seems clear by analogy that the same will be true of H-bomb testing. In the competition between offense and defense, versatility and ease of delivery are all-important. If anyone thinks that future tests are not important, let him consider how difficult it would be to convince either country to give them up without assurance that the other was doing the same. In the arms race, a unilateral cessation of H-bomb tests would, of course, be considered suicidal.

If both sides should discontinue H-bomb tests by agreement, both would presumably go on stockpiling the best H-bombs they know how to make, introducing as radical theoretical improvements as they might dare without tests, but the bombs in the stockpiles would gradually become more and more old-fashioned relative to what they would have been if the tests had continued. Advances in long-range rockets and other vehicles for delivery would have to contend with clumsy, old-fashioned warheads. Meanwhile, on the defense, development of radar screens and interceptor missiles would go on unhampered by the agreement.

It is impossible to predict just how much difference the test ban might make, but it is clear that the day of push-button destruction of cities by intercontinental rockets is not yet here, and this is one way to slow down or prevent its coming. It is a characteristic of long-range rockets that the payload is small compared to the rocket that takes off. H-bombs differ from A-bombs, it has been frequently said, in not being limited by any "critical mass." They may therefore be assumed to be very heavy, which may prevent them from being delivered by intercontinental rockets for some time, or perhaps permanently if they are not much further developed. The proposed H-bomb test ban could then, for all we know, have the effect of confining the delivery of H-bombs, or of all but the weakest H-bombs, to airplanes that will be much more easily intercepted than rockets. The effect of the test ban may be not so drastic as that, but it cannot help but be in the direction to favor defense over offense.

Monitoring the Test Ban

The unpleasant international incident accompanying an H-bomb test earlier this year shows that even in the vastness of the Pacific it is already difficult to find a proving ground adequate for the purpose. It is only a very minor advantage of the test ban proposal that it would get rid of these small but perhaps growing difficulties, but that incident of the Japanese fishermen does dramatize how easy it must be to detect H-bomb tests on an intercontinental basis. Even A-bombs are not only detected but extensively analyzed from a distance. President Truman, on September 23, 1949, announced, "We have evidence that within recent weeks an atomic explosion occurred in the USSR." Speaking later of that evidence,[2] Dr. Vannevar Bush said, "When we reviewed the evidence for the first Russian atomic explosion, we did not find out merely that they had made a bomb. We obtained a considerable amount of evidence as to the type of bomb and the way in which it was made." This seems to indicate that the detonation of practically any nuclear weapon can be reliably detected at long range.[3]

If it should perhaps be necessary to admit the possible immunity of very weak A-bombs, detonated under special conditions, from reliable long-range detection, this would not seem to be a very important flaw in the effectiveness of a test ban, and even with this possibility it would seem desirable to include A-bombs in the test ban. Besides retarding large A-bombs, the ban on A-bomb tests would have the special advantages of impeding the further development of the A-bomb trigger of H-bombs. There is one possible argument against including A-bombs in the test ban. The main advantage of the ban should be to favor defense over offense, and the possibility has been mentioned in the press that small A-bombs might be used in defensive interceptors, with the future development of which an A-bomb test ban might or might not seriously interfere. It is difficult to assess this point without further information, so we speak in this proposal mainly of an H-bomb test ban, although A-bombs should probably also be included.

The monitoring of an H-bomb test ban should be done by an international agency rather than simply by each of the principal competitors monitoring the other, in order that there can be no propaganda claims to doubt the detection by an opponent, perhaps accompanied by the charge that he is attempting to hide his own tests. The international monitoring agency could be kept above reasonable suspicion and would be able to announce unequivocally in what general region of the earth a violation had taken place. Its purpose would be to be ready to do this so that no violation would take place.

The international agency would not have to have access to any national territory where it is not wanted. It would probably be in the self-interest of the chief contracting parties to provide operational bases for the agency on their windward coasts, which would cover the northern hemisphere, and similar island and coastal bases would presumably be made available in the southern hemisphere as well, by various nations interested in promoting world stability. Monitoring operations could also be carried on at sea if necessary.

The Possibility of Violation

In making a critical examination of an arms limitation proposal, it is necessary to ask what would happen if there were a violation even though one expects the proposed agreement to function without violation, because the answer to this question is a measure of the effectiveness of the proposal in discouraging violation. Even a real disarmament agreement by treaty between sovereign powers, which is more ambitious than the test ban we have been discussing, really amounts only to a very early warning system. A potential aggressor knows that he can start an arms race in preparation for a war that he intends to start, or can even start the war with limited armaments, with the expectation that both sides will then start making all the weapons they can. The direct advantage of the disarmament agreement is that he is apt to be much less tempted to start a long-drawn-out war of which he cannot predict the outcome than to start one with what may seem to him an overpowering initial blow by the surprise use of unlimited weapons. There are also, of course, indirect advantages to a disarmament agreement, such as getting people in the habit of thinking about cooperating and permitting a larger share of the world's economic resources to be applied to constructive purposes.

With the very primitive type of arms limitation provided by the H-bomb test ban proposal, the direct advantage is, of course, not so great as with more substantial disarmament, but it is an advantage of the same sort. A violation would presumably mean that the unlimited arms race would be on again, tests and all. The test ban is proposed in order to reduce the attractiveness of sudden aggression, not for the sake of saving effort on armament. With the test ban, each side would in the interests of complete preparedness want to think what tests it would like to make were it not for the ban, so as to be ready to enter the unlimited arms race quickly if the other side should commit a violation. If each side is convinced that the other is not napping, neither is nearly as apt to declare aggressive intentions by violating the test ban before finishing the development and production of a crucial new weapon than it would be to make a surprise attack with the completed weapon. Aggression would thus seem much more inviting without the test ban.

Defense Vital in Discouraging Aggression

Let us look a little more closely at the question of comparing our future security with and without the proposed test ban. Our present policy puts all our trust in discouraging aggression by our power of massive retaliation. Influential extremists have gone so far as to proclaim that any diversion from our offensive effort to defensive preparation is tantamount to treason, but recently some hope has been expressed that a more nearly balanced view may have begun to hold some sway. There remains a very strong feeling that the more completely we are able to obliterate the Soviet Union, the more will we be able to deter aggression, no matter what capability they may have of obliterating us.

Incidentally, this residual feeling that a powerful offense should be developed at all costs is a politically powerful deterrent to our seeking any arms limitation agreement because it is consistent with our experience in the two world wars, an experience which manages to set itself up in many quarters as the criterion for true Americanism. With our splendid continental isolation which was our substitute for defense gone on the wings of advancing technology, that experience means exactly nothing.

Massive retaliation requires surviving the initial attack well enough to mount the retaliatory offensive. Anything more anticipatory than this is not retaliation but "preventive" attack, unless it be imagined that our timing can be so perfect as to be sure that the retaliatory armada will pass the attacking force in midair, subject to recall if the anticipated attack miscarries. Our policy of massive retaliation then requires some measure of defensive capabilities, or that the enemy attacking force be short of completely overpowering. The potential aggressor needs to be convinced not only that we have the offensive capability of penetrating his defenses and laying waste his land; it is just as important that he be convinced that he cannot forestall our retaliatory attack by a crippling surprise blow. Thus, even within the limited framework of our policy of massive retaliation, what he can do to us and in particular to our launching sites is quite as important as what we are prepared to do to him.

If we proceed with the unlimited arms race, both sides will presumably develop rapidly on toward the era of the complete "push-button war." We will develop the capability of delivering, after allowing for losses in penetrating the defenses, several times as much destruction as is required to destroy all important targets. We thus need be sure that only a modest fraction of our retaliatory force can escape the initial attack. On the other hand, the Russians will presumably also be capable of destroying our important installations several times over, so the initial attack is apt to be relentlessly thorough and it will perhaps not be likely that a sufficient retaliatory force will be missed and take off. The apparent desirability of making an attack will be continuously and sensitively subject to the whims of judgment concerning this delicate balance. Such a situation seems extremely explosive. The protection provided by the power of massive retaliation will be flimsy indeed, and if the deterrent fails the initial attack will be obliterating. If as a final limit we consider the time when both sides have completely devastating offensive capabilities, retaliation will be impossible after the first blow, the only security will lie in not allowing an opponent to strike first, and the premium on surprise attack will seem decisive.

Reliance on Retaliation

Our present policy of relying almost exclusively on our capability of massive retaliation as a deterrent to atomic attack, cogent as it is in this transition period at the beginning of the atomic age, is clearly only temporary in its effectiveness. It is a grim necessity to assure a reasonable degree of short-term security. The pace of technical developments makes it essential that we plan now also for long-range security. Lacking any clear solution of the long-range problem, the best we can hope to do is to slow down the worldwide pace of development of new techniques of destruction and thereby extend the time during which the threat of retaliation acts as a deterrent to aggression. If it were technically possible to accelerate worldwide defensive weapon development to such an extent that an almost 100 percent defense against atomic attack would become possible, this would constitute a solution of the long-range security problem, but unfortunately this is not to be anticipated. Short of this, a decided trend favoring the more rapid development of defense than offense would serve mainly to extend the time during which a retaliatory policy may help prevent war.

During the first decade of the atomic age, when we were confident of having undisputed numerical superiority in atomic preparations and of a Soviet stockpile still incapable of devastating our country completely, we could feel fairly sure of our ability to launch retaliation after a first blow. Now that we are entering the stage of nascent saturation, in which numerical superiority of striking power begins to lose its significance, it is important that we do not allow ourselves to be misled by habits of thought developed during the earlier period. Quite contrary to some earlier and still popular impressions, creating the conditions to favor defense from now on increases and prolongs the effectiveness of the policy of massive retaliation. Rather than rejecting the test ban proposal as somehow antithetical to the policy of massive retaliation, as has apparently been done in the past, we should now consider it an indispensable adjunct of that method of minimizing the chance of all-out war.

Test Ban Proposal Not New

The release by the AEC on June 15 of the testimony in the Oppenheimer security clearance procedure provides some new information by which we may judge the technicalities of the world atomic situation and some of the reasons for the formation of present national policy. The testimony of Dr. Vannevar Bush[4] suggests that the H-bomb test ban proposal was considered and rejected within the high councils of our government even before the fateful test that marked the beginning of the thermonuclear age. Then a test ban agreement could have been much more effective than it can be now, for it would then have essentially confined the stockpiling to A-bombs. Now that we know that the first Soviet H-bomb test came only nine months after ours, it seems in "hindsight" a pity that the H-bomb test ban was not pushed in diplomatic channels before those tests. It might have been accepted and the worst that could have happened would have been delay enough to lose us our nine months' lead with an undeliverable weapon which could hardly have been crucial if our numerical superiority in deliverable and enormously powerful A-bombs was as great as supposed. At the time it was, however, not clear that the worldwide technological advances were so rapid, and it was hoped that if the test were successful we might this time retain our monopoly for a long time as an overpowering instrument of our policy of massive retaliation. That we missed that opportunity through the exigencies of development and perhaps through faulty evaluation of the situation in the past is not a reason why we should not retrieve as much of its advantage as we can now that we are aware of being well on our way into the thermonuclear age.

The testimony also tells of the "very brilliant discovery" of 1951, and makes one wonder if perhaps the H-bomb, like Athena, did not "spring full-grown from the brain of"—well—Teller (and someone on the other side of the world). The fact that the A-bomb has ingeniously exploited the heavy end, and the H-bomb the light end, of the periodic table of elements with its bow-shaped graph of mass defects, also makes one wonder if instruments of mass destruction may not already have reached final perfection. If so, it would be surprising, almost unprecedented. One could have wondered the same thing about the steam engine a century or more ago, but the efficiency and compactness of power generation from fossil fuels has increased ever since. Though this point may bear classified investigation, we had best not pass up the urgent present opportunity of the H-bomb test ban proposal under any general delusion that there will be no further development.

Effects Gradual and Fair to Both Sides

Not only does the test ban proposal hold great long-range promise for greater world stability because of its gradual reduction of the preponderance of offense over defense, but it should be acceptable also because it does not interfere with plans for short-term security, its effects being so gradual. It leaves armed might and the threat of retaliation as the principal shield, with gradually increasing emphasis on continental defense. It permits both sides to go on stockpiling H-bombs which in later years will become clumsier and harder to deliver than they otherwise would have been. It treats both sides alike, and while putting a gentle lid on their *absolute* military strengths leaves their anticipated *relative* strengths, which is the important consideration, approximately unchanged. It requires no drastic alteration in our attitude toward international relations, nor even any settlement of bones of contention in the cold war. It requires no relinquishing of national sovereignty, save only the right to conduct H-bomb tests. Its value does not depend on anticipating any further agreement but it does not stand in the way of, and indeed provides more time to try to reach, a more ambitious arms limitation agreement and settlement of differences in other fields.

The self-policing nature of the test ban proposal makes it entirely different from anything we have yet proposed to the USSR for it allows them to keep their cherished Iron Curtain, and our frustrating past experience is not adequate reason to believe that the plan if properly presented would not be acceptable to the Soviet leaders. They, too, stand to gain by an increase in the overall effectiveness of defense if they hope to go on trying to achieve their aims without precipitating a world atomic holocaust.

There is another argument which should make the H-bomb test ban proposal attractive to the two present H-bomb powers, and that is that it will prevent the H-bomb race from becoming a many-sided affair, for no other powers can independently develop H-bombs without making tests. What little has been told about the techniques of making H-bombs indicates that they are much easier to produce than was anticipated. This would appear to make the rise of other H-bomb powers rather imminent. In the era of traditional weapons, a "third force" was considered a stabilizing influence because it could by careful statesmanship maintain a balance of power and perhaps prolong the peace. But with the tremendous premium on surprise attack provided by atomic weapons, although the magnitude of the prospective retaliation is a strong *rational* deterrent, the danger is that some statesman will act irrationally as statesmen have been known to do in the past, and it takes only one mistake to destroy civilization as we know it. The rise of more H-bomb powers will have as its main effect the introduction of more statesmen with the power to make the fatal mistake. If the statesmen of each of the two present contending powers feel that they will on their side not start the war and there is a good chance that they can play the game in such a way that the other side will not for quite a long time, let them look together to this means of keeping it from becoming a complicated triangle game. As for the other powers, they in their present impotence would probably be glad to subscribe to any agreement that will in some measure clip the wings of the two "atomic colossi" and make for a more stable world.

If there is any hope left among responsible statesmen, this plan for a small but significant and practical measure of arms limitation, the H-bomb test ban proposal, should have much more insistent promotion than it has yet received.

NOTES

1. The Soviet position advanced favorably when Vishinsky, before the UN Assembly on September 30, subscribed to a French-British proposal of last June, which, while sketchy, provides for very substantial inspection and control. Despite the possibility of future backsliding, this concession brings a new opportunity in the search for avoiding the grim H-bomb threat, if only our negotiators can rise above the country's accusing mood and make a sincere effort.

2. Transcript of Oppenheimer hearings, p. 564.

3. James R. Arnold, "Tracing Nuclear Explosions," *Bulletin of the Atomic Scientists* 9 (March 1953): 44–45.

4. *New York Times,* June 17, 1954, p. 22; transcript of Oppenheimer hearings, p. 562.

Why I Am for Stevenson

[1956]

I am for Stevenson because I am deeply troubled about the threat of atomic annihilation.

The almost unimaginable destructiveness of hydrogen weapons puts us in a crucial predicament that most of us refuse to recognize. The deterrent they provide is effective only so long as the decisions of national leaders are entirely rational. We are in the midst of an ever mounting arms race of which we can foresee no end, save the disastrous one. How to taper off the arms race poses an enormously difficult problem that demands imaginative thought, unstinted research, and determination to find a prudent solution even if it requires departure from traditional attitudes. I am for Stevenson because I believe he understands and feels this as no president has.

The problem is a hard one to discuss in an election campaign because it is intricate and partly hidden in secrecy. I am for Stevenson because his boldness in introducing the issue into the primary campaign and the imaginative spirit of the proposal he there made show both his burning interest in this problem and his willingness to consider seriously a real change of national policy. He may not stick to just his present proposal after learning, as president, all the secret details of various facets of national interest. But if it does not stand up, I believe he will insist on a real search for something better.

He proposed that we should short-circuit the protracted negotiations by taking a decisive initiative in the simplest safe step in arms limitation: we should stop further testing of H-bombs and announce that we will not start again unless some other nation goes on testing them. Such a cessation of tests would seriously impede the development of new horror weapons about equally on both sides, and is the one type of arms limitation that can be adequately monitored without area inspection. This direct approach is a way to take the Soviets up on their boast that they favor ceasing tests. If they should join us in ceasing tests, as seems rather likely, the de facto cessation would facilitate getting an international agreement against all nuclear tests. If the Soviets should ignore our bid and flaunt world opinion by continuing their tests, we would resume ours, having delayed our testing, but not all our development, by a few months.

Judged from the outside, this proposal makes good sense. Not so President Eisenhower's rejoinder that there would then be no purpose to developing ICBMs without further developing their hydrogen warheads. This is an interesting disclosure of the state of the art but completely missed Stevenson's point that we should confine ourselves to stopping only those developments that we can be sure the Soviets are also stopping.

In spite of slight misgivings, lingering from my Republican past, about some aspects of his domestic program, I am strongly for Stevenson because I believe the arms race raises by far the most crucial issue facing us today, and the present administration has not adequately tackled it. Our immediate military needs for effective deterrence have been vigilantly pushed, but there has not been adequate concern for the future. To depart from traditional preatomic attitudes to meet this unprecedented challenge would require forthright decision by an inspired leader, but the question appears instead to have been delegated to the administrative team. One member of the team, Governor Harold Stassen, has devoted himself valiantly to it of late, but even he has been handicapped in leading long-range studies by the burden of responsibility for the immediate international negotiations, which has placed him under the step-by-step scrutiny and veto of other departments.

The result of all this indecision by teamwork has been that we have been unable in the negotiations to break away from the time-honored procedure of making offers so slanted that the other side was almost sure not to accept. One of these was the famous ''open skies'' proposal, which would be a fine start if we could arrange it but had little chance of acceptance because the Soviets have been able to hide so much from us by their Iron Curtain, and the quid pro quo was lacking. We were not even prepared to explore their intentions and try to exploit the possibilities when last spring they proposed the establishment of ground check points for inspection activities throughout large countries.

It is not enough to point to mutual deterrence as sufficient safety for the present and to let the future take care of itself. The revolutionary technical developments now in progress cannot be quietly undone by a future generation. The arms race is rapidly getting more difficult to bring under control while we continue without a national policy to try to control it, even as Rome burned while Nero fiddled. I am for Stevenson because I believe he won't fiddle.

Evasion of the H-Bomb Issue

[1956]

President Eisenhower early in the campaign answered Adlai Stevenson's suggestion that H-bomb tests be stopped by quickly reviewing administration policy on arms development and falsely implying that "atoms for peace" is somehow a relevant accomplishment. Failure to have developed a vigorous policy to meet the future threat of the endless arms race was hidden by expressing the impracticable desire for actual disarmament with the guarantee of adequate inspection. The president then rested his case—he had said his "last word." A subsequent administration statement, the so-called White Paper, of October 23, shows that the president kept his original promise. He had indeed said his last word. For in this paper, it is just said again. The statement does, however, contain some interesting bits of clarification.

> If we were to suspend research and preparation for tests . . . we could find our commanding position in the field erased or even reversed. The preparation of such tests may require up to two years. [Further] If we suspend only the tests, while continuing precautionary research and preparation . . . we could suffer a serious military disadvantage. It requires a year or more to organize such tests as those conducted at our proving ground in the Pacific Ocean.

This last estimate deserves careful scrutiny—it is perhaps the only solid new material in the report. "Precautionary research and preparation" comes before the year required. It includes conception, design, and construction of the nuclear weapon that would be tested if testing were permitted. To "organize and effect such tests" alone would require a year, the report says. However, in view of the enormous importance of starting to make progress in the field of international arrangements to catch up to the devastating pace of weapons development, it would seem entirely practical to go to the expense of the further precaution to "organize" the tests ahead of time. This would include stockpiling and to some extent installing test apparatus, keeping a fleet of ships in readiness, and assigning personnel for possible quick transfer from other activities.

What we really need, then, for a fair assessment of the Stevenson proposal to stop H-bomb tests is an estimate of the time required for the last part, to "effect such tests." A reasonable estimate would be about three months to mount and carry out the tests, *if they were organized on a standby basis ahead of time.* The administration might estimate a bit more. Someone apparently has estimated half that time, but three months would seem to make sufficient allowance for human imperfection and for expending less than the maximum useful preparatory effort.

The conclusion supports Stevenson's contention that to take the initiative and even without formal agreement stop H-bomb tests and challenge others to do likewise would risk only a very few months' delay in our testing program and even less in development. In view of Eisenhower's statement that with two years' delay in testing and development "we *could* find our commanding position in the field erased or even reversed," a delay of two or three months would not be crucial. It would be a slight sacrifice, to be sure, but think what it would buy!

There are two possibilities. Either the Russians do or do not want to join us on the path of guarded mutual concessions in order to try to avoid blowing each other to bits. They are having outstanding propaganda success these days by claiming that they do, and we've lost a lot of friends by our inflexible attitude. Suppose they don't want to join us. Then we lose two or three months of testing and win back friends throughout the world.

Suppose instead that they do join us. Here is the possibility of the really big gain. We would make a major departure from the headlong race in unlimited arms development for which we can see no end but disaster. The arms race would go on, the stockpiling, some aspects of development, but the rate of development would be limited. President Eisenhower informed us last spring, again in answer to Stevenson, that further testing is necessary to develop the H-bomb warheads of intercontinental ballistic missiles. It will probably be a fairly long time before these weapons have sufficient accuracy to be effective without area incineration by large H-bombs. To slow down the development of these future Russian instruments of instantaneous devastation, along with our own, would itself be an important advance. It would help keep warning times for sneak attack in hours rather than minutes or seconds. Beyond this test cessation the future possibilities are boundless, and we would have bought time and established a climate for working them out.

If we continue instead with our present policy, the future possibilities are narrowly circumscribed. We can do nothing but follow the one-way street of the limitless arms race to its bitter end. If we pass up this propitious time, there will be no turning back, for weapons will become too insidious, too instantaneous, for control. There will be many nations in the race. We may avoid having anyone make a mistake for a time. Our "commanding position,'' as the president calls it, may be expected to be influential for the present. But the greater danger in the long run is not that Russia will make a rational decision to challenge our strength, but that someone—perhaps some local commander in Russia or the Argentine with his finger on the push button of swift havoc—will make an irrational mistake. Other human institutions leave some margin for error. This one does not. One mistake will mean the end for all.

This is the issue that the president deliberately evades in his campaign. He is guilty of seriously misleading the American people. He and others of the administration have steadfastly pointed to the dangers, such as the time lost in development, in stopping H-bomb tests. They have consistently refused to recognize that there is any danger in following our present inflexible course. Mr. Eisenhower has further confused the issue with his incessant implication that "atoms for peace" is a substitute for armament limitation. It should be called "atoms for industry" because it is merely a type of economic aid and has nothing further to do with the prevention of war. He has confused the issue by claiming that the "open skies" proposal was a bold move in the direction of arms limitation, when in reality it was a timid move that never had much chance of acceptance. He seems proudly content to have made a feeble try and failed.

Most voters lack the time to grasp the intricacies of the problem and may not easily recognize the utter failure of the administration team to agree on a forthright policy in pursuit of future peace. Convincing testimony is found in the transparent evasiveness of the president's argument, the hammering on "atoms for peace," the pride in the "open skies" failure.

To follow indefinitely the present uncompromising course of the unlimited arms race is to invite, almost to guarantee, eventual disaster. We need a leader with the courage and the prudence to lead on the long and difficult road in the opposite direction. By vigilantly maintaining deterrence with our present types of weapons on both sides while tapering off the development of new weapons of ourselves and our rivals, we will have less danger of a flareup on the way and a tenable goal for the future.

If We Just Go On Testing . . .

[1960]

Give those who distrust an agreement banning nuclear tests the thread of an idea and they'll make a rope out of it with which to hang the Geneva negotiations. An article in this month's *Foreign Affairs* by Professor F. J. Dyson is illustrative. Mr. Dyson spins a gossamer theory: Radically new and cheaper nuclear weapons are possible; the testing of such weapons in very small sizes is undetectable; we should hesitate to deny ourselves these weapons by agreeing to ban further tests.

What is this theory which Dyson is advancing as a new argument for continuing our nuclear testing program? The new small weapon he foresees is based on the hope of attaining the fusion of light elements directly from heating them by chemical explosives, without the help of the fission booster which made possible the H-bomb. In support of this hope, he offers just two technical facts, neither of which says anything at all about the practical likelihood of such a development. The first fact is that, "There *seems to be* [italics added] no law of nature forbidding the construction of fission-free bombs." This is the same as saying we can't prove definitely that they can't be made some day. The second fact is that neutrons have been produced by heating and squeezing heavy hydrogen with chemical explosives. This may sound impressive to those unaware of the enormous difference between producing a few neutrons and making a thermal chain reaction; it actually means no more than to say that because a man is healthy enough to get up out of bed, he can climb Mt. Everest.

The main point is that it was far from easy to achieve an H-bomb, even with the benefit of a fission booster giving a millionfold boost beyond chemical energies, and it seems much too much to expect to make up for the lack of so big a boost by cleverness alone. People who expect another scientific "miracle" simply because they have become accustomed to "miracles" should realize that nature gave us two favorable energy trends at the two ends of the list of elements: one at the uranium end for the A-bomb, the other at the hydrogen end for the H-bomb; and there aren't any more. Dyson's notion seems like a very long shot indeed.

His enthusiasm over the remote chance that a cheaper small nuclear weapon might just possibly be developed is hardly enough to justify his conclusion that we should only consider a test ban that is perfectly evasion-proof. In taking this position, he ignores the threat of accidental war; he seems to expect the world to go on with the mounting tensions of an expanding arms race proliferating in radically new weapons; and then, suddenly and without preparation, he anticipates that all nations someday will have the political maturity to "hand over all such devilish inventions to an international authority powerful enough to prevent their abuse." It is a philosophy of crisp concepts: a perfect test ban or none at all; completely uncontrolled chaos or complete world order; fission-free bombs either demonstrably impossible or very probable.

A gradual approach is needed if we are to evolve workable international agreements. It would be more difficult to control these "devilish inventions" now than it is to monitor a test ban to limit their further development, and it will become increasingly difficult to control them if testing continues to spread them throughout the world and make them still more "devilish." A well-monitored test ban holds the most promise of reducing the likelihood of war. We must not be deterred from this quest by contrived roadblocks like the recent exaggerated publicity for geophysical experiments with A-bombs or Dyson's blown-up fears of a long-shot development which, even if accomplished, would not be crucial to a military posture ultimately dependent on the "great deterrent."

At this moment, there is a chance, perhaps a last chance, to start to bring the testing of nuclear weapons under some kind of reasonable control. There are still parts of a test ban treaty on which the U.S. and the Soviet negotiators have not been able to agree, particularly the question of the number of on-site inspections that would be permitted and the voting procedure to be followed in the control commission, and these matters remain a challenge to our bargaining astuteness. But what counts most now is getting on with an agreement permanently banning all tests above the "twenty-kiloton limit" and installing as soon as possible a control system of monitoring this effectively. At the same time, we must be willing to agree to an only partially monitored moratorium on smaller tests for two years or more, and get to work on joint research to improve detection techniques for these small-yield blasts.

The Rest of the Test Ban

[1963]

Now we have part of a test ban. It's worth a great deal to have made a first formal, legislative step in deemphasizing the arms race. It's worth something to be rid of the unpleasantness of radioactive fallout from tests, even though this in itself has little bearing on the likelihood of nuclear war.

Although wary about small evasions of an underground test ban, the West has been proposing an aboveground test ban since 1959. American and world public opinion has been clamoring for an end of the fallout from atmospheric tests. Until this summer, the Soviets held out for a complete test ban, and it seemed that the price of satisfying the public clamor would be for us to swallow our exaggerated fears of underground evasion, and muster some faith in our massive nuclear superiority or in its meaninglessness in the light of overkill capabilities.

Valuable though a complete test ban would be for the sake of impeding the spread of nuclear weapons and reducing the likelihood of nuclear war, the Senate seemed reluctant to take this step. Yet the force of the clamor, together with the administration's appreciation of the real need for a complete test ban, made it seem almost probable that the president would take the plunge and with great effort would be able to gather enough votes for ratification. Largely because of the clamor against fallout, the Senate might have paid the price without fully appreciating the value of what it was getting.

This bubble was burst by Khrushchev when he made us the gift of a partial test ban, relatively easy to ratify. In a sense, the politically priceless clamor against fallout has been squandered to buy a lesser bauble, to treat the symptom rather than the disease. Now we're faced with the problem of finding the political wherewithal to buy the rest of the complete test ban that we'll probably really need to make much progress on the Nth nation problem.

Before we cry over this spilt milk, it is best to recollect that ratification of a complete test ban this year seemed at best rather doubtful. We may, then, rejoice that half a test ban in the hand is better than a whole one in the bush, and explore how to go on from here.

The test ban debate in the Senate was notable for the extent to which the senators felt compelled to show themselves before their constituents as concerned primarily for security through superior armament. Though not all senators were convinced, assurances were emphasized that we are far ahead of the Soviets, that the ban on aboveground tests will leave us so, that we will increase our activity in underground testing to compensate for the uncertainties of the ban, and that we will be ready for resumed atmospheric testing in the event of abrogation of the treaty. Much of this makes sense as fact and prospective behavior under a partial test ban, but the need for such emphasis in order to "sell" it to the Senate is disturbing when one contemplates the chances of obtaining ratification of further steps on the road to disarmament. In particular, it makes sense to let a partial test ban limit us only where it applies, to be ready for resumption in the event of abrogation at least until experience with the ban gives us further confidence in its permanence, and to learn what we can of military value from underground tests until we manage to ban them, too. Experience will probably show how little of military value advanced nuclear nations can learn from underground tests and will improve our knowledge of how to detect them. If so, such tests may either lapse from mutual uninterest, or more likely, a ban on them will be more easily approved by the Senate on the basis of unobtrusive methods of control. It would probably take several years for such a trend to run its full course, but before then the trend may be helpful if other pressures for a complete test ban are strong enough.

The Nth nation threat may apply this pressure in two ways. First, a new nuclear nation can learn more from underground testing than an advanced nuclear nation can—it can prove the power of its new weapons. Testing underground is expensive compared to that aboveground, but cheap when compared with the cost of a bomb development program. Although a nation adhering to the partial test ban has less incentive to invest in bombs if it can't test their effects in the atmosphere, it might still decide to go ahead on the basis of underground tests. If the prospect of such decisions develops, a ban on all tests will be more urgently needed.

A second source of pressure may be provided by the reluctance of certain recalcitrant nations to become involved in a test ban unless it is a complete one. De Gaulle seems to go still further and wants it coupled with far-reaching disarmament before professing interest. With only a partial test ban, the nuclear "haves" are asking a great deal of restraint by the "have-nots" in exchange for a minimum of restraint by themselves. While the less developed nations should welcome this restraint as a contribution to reducing the likelihood of war and should be happy to cooperate as so many already are, those that are nearly ready to take the plunge may well hold out for more restraint by the giants.

While the test ban agreement has a definite value in itself and as a harbinger of other steps, some tend to overrate it. Tristram Coffin, for example (in the July–August issue of the *Correspondent*), sees the sudden breakthrough to the partial test ban as evidence that both East and West have concluded that the arms race is too expensive and dangerous. This may be true of the East, but the militarily overcautious and almost bellicose undertones of the Senate debate, the unprecedented rate of missile production this year, and the fact that the treaty results from acceptance of our four-year-old proposal and is so mild as to leave the arms race in full swing, all suggest that this diagnosis of the West is premature. The hope is rather that the change of mood will set the stage for such a conclusion. It was gratifying to observe in the test ban debate that some senators were seriously grappling, perhaps for the first time, with the insecurity that comes with the arms race and some were eloquent in expressing awareness of the long-term dangers.

One of the motivations of the arms race has been for each side thereby to try to break the economic back of the other. Despite our greater productivity, the Soviets may have aspired to do so because of the multiplication factor possible with their secrecy while we, not knowing of the multiplication factor with certainty, have perhaps thought we were inducing greater Soviet expenditures than we were. This was true of heavy bombers. The Soviets practically skipped the heavy bomber stage and by parading a few impressive prototypes induced us to build a massive fleet. It may be true also, though probably to a lesser extent, of ICBMs. The Soviets' change of heart on the partial test ban may indicate their recognition that our economy can stand the strain, and that there is a limit to the factor of superiority they can safely permit us and still maintain deterrence, even after discounting for overkill. If, with the change of mood, we become more aware of the adverse multiplication factor against which we are working on the Soviet economy, we may be more willing to give up the economic exhaustion game as not worth the candle. If judged solely as a deterrent, the magnitude of the superiority ratio we have been maintaining might then come to seem unnecessary.

The change of mood may give us great confidence that the Soviets are not hankering to launch an attack on Europe. A nuclear free zone in Europe could result, particularly if this confidence should be bolstered by a first-stage disarmament exchange in which Soviet superiority in heavy tanks is pared down in exchange for some reduction of our nuclear superiority. With the difficulties being encountered in designing a command and control for a European finger on somebody's nuclear trigger, such a course might well come as a relief even to the military planners.

A substantial reduction of Soviet conventional striking power, through the destruction of tanks and similar measures, could reduce our dependence on the West German divisions in NATO. This should permit us to abandon our policy of insistence on the reunification of the country that started the two world wars, a policy of dubious merit and impractical revisionism which seems to stand squarely in the way of meaningful disarmament agreement.

Among thinkers on arms limitation there has been contention between those favoring limitation by formal agreement and those preferring reciprocated unilateral acts. The partial test ban is a first step in formal agreement, following some reciprocated unilateral restraint, particularly in backing down from Cuba. Some easing off of arms expenditures, initiated by the Soviets who have particular need of it for consumer goods and reciprocated by us, might be the next important step in reciprocated unilateralism. It might start us working seriously on the economics of the conversion to disarmament. While such reciprocated unilateral steps may be the easiest ones to take next, we must remember that the partial test ban has shown the greater effectiveness of negotiated agreement for accomplishing certain critical ends. The test ban will do what a moratorium could not.

While prepared to rejoice in any reciprocated unilateral steps that may ease tensions, we must keep attention focused on the need for further formal agreements on the road to substantial disarmament. High on the list should be the rest of the test ban as one important element in convincing the nonnuclear nations that their best interest is served in not contributing to the further proliferation of nuclear weapons.

3 Proliferation

It is greatly to the advantage of both of the major nuclear powers to have nuclear weapons capability limited to as few other nations as possible. This was clear early in the arms race, as indicated in the following articles, which also discuss how a nuclear test ban would be a useful means to that end. It is also indirectly to the advantage of the countries without nuclear weapons, since it makes less likely the worldwide calamity of nuclear war. That is a long-term view, but some nonnuclear weapon states may be more influenced by the short-term incentive to acquire nuclear weapons before their neighbors do. Thus international pressures are needed to dissuade them from going nuclear.

The Nonproliferation Treaty (NPT) of 1956 was intended to supply such pressure. If all nations would sign, it would provide each with assurance that its neighbors were not going nuclear. It incidentally made provisions for the more advanced nuclear nations to help others with the technology for nuclear power and with the anticipated peaceful uses of nuclear explosives, a passing fad of the time that was promoted by opponents of a test ban (discussed in the article "Special Report on Plowshare"; see supplemental papers list at the end of this book). However, the treaty demanded that the nonnuclear weapons states give up some of their cherished national sovereignty, the freedom to make these powerful weapons, with no corresponding sacrifice by the nuclear weapons states. The best they could get from the nuclear weapons states in return was the promise, in Article VI of the NPT, to "negotiate in good faith" on steps of arms control. This was a formal expression of the incentive the nuclear giants should feel to get on with arms control while convincing others to remain nonnuclear.

It seems remarkable and disappointing that, in all the years since the papers of this chapter were written and since that treaty was signed, the nuclear giants, though they have negotiated, probably at times in good faith, have not lived up to their end of the implied bargain even to the modest extent of achieving a comprehensive test ban. In the light of this it is perhaps not surprising that only about two-thirds of the nations of the world have adhered to the treaty. It seems remarkable and somewhat gratifying that, in spite of all that, only two more nations, France and China, have acquired substantial nuclear weapons capabilities.

Nuclear deterrence between two major powers remains stable as long as the leaders of both powers act rationally and as long as there is no operational accident to start a nuclear war unintentionally. It is a tribute to the perfection of the operational procedures of the present nuclear powers that there has been no accidental detonation of a nuclear weapon, particularly of one aimed irretrievably at another country that might start a nuclear war. People are fond of quoting Murphy's Law to the effect that anything that can go wrong will, yet of a total of some fifty thousand nuclear warheads in the arsenals of the world, many of them on instant alert for years, none has exploded accidentally. This record is much more impressive now than it was at the time these papers were written. It has been achieved not without exercise of extreme precautions by the United States, and presumably likewise by the USSR. If one person at the controls could launch a missile, a real danger would be that he might go mad with the tedium of being constantly on the alert, year in and year out, without any action. One precaution early in the intercontinental ballistic missile (ICBM) era was to have firing depend on concerted action of two officers with two keys. The precautions in U.S. practice have since become much more sophisticated. No one at a local control post (except in submarines!) can fire a missile without receiving a secret code from headquarters along with the order to launch.

If many nations with less extensive technical resources were to possess nuclear weapons, it seems unlikely that they would all develop such sophisticated precautions, and the possibilities for accidental nuclear war would be increased.

This technical source of trouble, while serious, is perhaps not as serious as the political one. There have been many hotheaded acts by leaders of third world countries and even more by terrorist groups in recent years. The prospect of nuclear weapons in the hands of some of them is indeed most alarming.

Proliferation has not proceeded as fast as was anticipated in these papers, but other nations are coming close, and avoiding further proliferation is still a powerful incentive for achieving a comprehensive test ban, for making real progress in other measures of arms control, and for limiting the availability of nuclear weapons materials and techniques.

Much has been made of the fact that nuclear energy can serve us as well as destroy us. To my mind the greatest of the dangers in developing it as a useful servant is the likelihood that, in so doing, we are spreading its availability as the Great Destroyer.

A group planning to make a nuclear explosive faces two main problems: obtaining the fissionable material—plutonium or highly enriched uranium— and developing the design and techniques for using it. The claim that a bright high school student can do the designing of a crude but still terrible bomb to be built in someone's garage is a bit overdrawn but not by much. A sophisticated laboratory such as might belong to a nation, but not to a terrorist group, would be required to design a very efficient A-bomb or an H-bomb. In either case the crucial materials that can be derived from the nuclear power industry are essential and experience in industrial nuclear technology could be a useful introduction to bomb making. A contribution, however small, to the likelihood of a catastrophic nuclear war is thus to be counted as an important part of the cost of proceeding with the nuclear power industry and spreading it to the far corners of the earth. This was a consideration in deciding whether or not to proceed with commercial nuclear power, as may be seen in the first article of chapter 1. The drive for commercial power, with plutonium as a by-product, prevailed over such qualms when the decision was made to pursue "atoms for peace" in the mid-fifties. In order to discourage diversion of critical materials from power reactors to clandestine bomb making, occasional inspection is required of the reactors in countries adhering to the Nonproliferation Treaty, but the inspection procedures are perfunctory and not very convincing. India has not adhered to the treaty and made its first nuclear explosive from materials supplied for a nuclear power plant.

Now that this cat is out of the bag and mankind is receiving some short-term benefit from nuclear power with too little regard for the longer-term danger, one way to mitigate the danger and direct us to a safer path is to promote alternative energy sources. Conservation to reduce waste can help but alone is not enough. With too-rapidly growing and increasingly industrialized global population, increased power sources that do not greatly strain the fragile environment will be sorely needed. With this in mind, my 1973 book on nuclear energy and its problems was followed by my 1978 book *Wind Power and Other Energy Options*, a subject that is pursued further in chapter 8.

National Security with the Arms Race Limited

[1956]

National security has ceased to exist in any absolute sense. Our awakening tomorrow to participate in the life of a happy and healthy nation is subject to the whim of a foreign totalitarian regime—a regime armed with H-bombs and presumably with the means to deliver them almost unhindered. That whim is fortunately kept in check by our ability to retaliate, giving us a perhaps illusory feeling of confidence that we will work and enjoy many tomorrows. But national security has been reduced to a question of probability, instead of certainty, that we will survive any year without being blown to bits.

The scale of the potential disaster is so vast that our foremost national concern should be to minimize the probability of devastating war, while preserving the essential freedoms we cherish. It has often been said that the new technical realities require an entirely new approach to statecraft, but the world has remained insufficiently influenced by these new dimensions of destruction. Political evolution does not seem to have been hastened by the hot breath of technical advance.

Though there has been no development of new forms for maintaining peace, there has been an encouraging toughening of the diplomatic skin. This manifests itself in a new restraint in reacting to border incidents, such as the shooting down of a plane, that formerly might have been considered a *casus belli,* and leads to a hope that the stalemate may keep its precarious stability for a long time, perhaps much longer than the time since the Korean war. But toleration of minor offenses can go only so far before it encourages larger adventures. The danger of small wars can hardly be overlooked in view of the military buildup taking place in North Korea and elsewhere. And if a small war should start, there is always a probability, at least, that it may grow into the dreaded big one.

While the mutual deterrence of stalemate has an apparent stability that seems gratifying for the short term, it must not be allowed to leave us complacent over the prospects for the long-term survival of civilization. The existence of some probability of failure in any one year adds up to an enormous danger in the course of many decades.

Such concern for the future—for our own old age and the lifetimes of our children—is normally shrugged off with the thought that things will be so different in the future that some way will then be found to stabilize the world.

The tragedy is that through our astounding and unabated technical developments in this generation, we are making it utterly impossible for future generations to bring the ever new terror under control unless they can accomplish political utopias far beyond our dreams, while we in this generation remain apparently unwilling to take the slightest step toward an international political evolution.

Such a blanket indictment of our generation is of course unfair unless we point out that some nations have seemed more reluctant than others to discuss the possibilities of progress in good faith. But the fact remains that there has been no progress toward disarmament or arms limitation. There has been no enthusiasm for seeking substantial agreements since 1946, and the recent London talks ended, as did their predecessors, in discouraging failure.

The problem of any international armament limitation or control is already so very difficult from the technical point of view that tenable objectives in this direction are severely circumscribed. It is generally regarded as impossible to detect nuclear weapons once they have been produced and hidden. The fact that many already exist makes an agreement to retire them from national weapons stockpiles unenforceable.

There are benevolent organizations, here and abroad, promoting universal, enforced, and complete disarmament of all weapons and all nations down to the level of police forces. It must now be recognized that this could be achieved only under the protection of a worldwide, supranational government armed with nuclear weapons—numerous and mobile enough to cover all possibilities of secreted national arsenals. In the present sad state of international distrust, there is only a faint hope of achieving such a step in political evolution. It is, therefore, necessary in the present world climate to examine less drastic palliatives. Yet, from the premise of the enormous danger which hangs on the slender thread of mutual deterrence, it must be admitted that the logical conclusion is the search for supranational government. Statesmen continue to do lip service to hopes for a happier future in which the arms race may be brought to an end, either through agreement between sovereign nations or through peaceful establishment of a world police power. Yet the prospect is that neither course will be technically possible if weapons continue to be developed at their present alarming rate.

The difficulties of establishing confidence in a supranational government, permitting acceptance of its authority as a substitute for the sovereign right to wage war, now seem formidable indeed. Even in a fancied future time of easier trust it is hard to imagine that the transition to such a government could be satisfactorily arranged if nations were equipped with the concealed apparatus of "push-button warfare"—the capability of destroying any other nation at will. If, on the other hand, it should, in that friendlier time, still be necessary to mount the relatively cumbersome offensive of present weapons to destroy another country, the establishment of the authority of a world army would remain entirely possible. While our present H-bombs have a ghastly power and cannot reliably be ferreted out, the planes or submarines that must carry them could be controlled, leaving no question of the supremacy of the world armed force.

The alternative way to end the arms race is agreement between sovereign states to limit armaments and eventually reduce their scale to safer levels, guaranteed by mutual inspection and coupled with various political and economic arrangements to make aggression unattractive, it not impossible. But again it does not seem that any such agreement can ever be made once the "push-button" stage is reached, with the ICBMs poised in inconspicuous holes in the ground beyond the reach of inspectors.

Preserve the Possibility of Agreement

Here, then, is a very real and tangible incentive for trying to limit the development of weapons of mass destruction (including their delivery vehicles): i.e., to save the technical possibility of an agreement ending the arms race for a later time when broad agreement may be politically feasible. We speak of uniform limitation all over the world, not of unilateral limitation by any one power.

This is a real incentive, but perhaps it is not tangible enough, immediate enough, to compete with the frantic needs of the arms race for our immediate competitive national security. There is another, more impelling and less virtuous, incentive.

Prevent the Arms Race from Becoming Many-Sided

The deterrence of stalemated armaments is a logical deterrent. So long as each side acts sensibly, we should be safe from the scourge of modern war. The danger is that someone will make a mistake. The United States and the USSR are living on a certain respect for one another's sanity. Each feels some confidence that, with somewhat restrained diplomacy and the threat of obliterating retaliation, a surprise blow can be avoided.

It is a diplomat's nightmare to think of trying to keep a stable world with dozens of nuclear powers. With so many sources of the spark to touch off the holocaust, with so many statesmen in a position to make the final error, it would be a miracle indeed if a decade could pass without disaster. Yet this ugly day of many nuclear powers will surely come if we continue to follow the conventional course of unlimited rivalry between sovereign nations.

Here, then, is the common motivation to an agreement that could increase the national security of both great powers—an agreement to prevent, or effectively retard, the rise of other nuclear powers without upsetting the balance called stalemate. It is often said that we can only trust international agreements, particularly with the USSR, to be honored so long as they remain in the selfish interests of the contracting powers. This, then, is the basis for an effective agreement.

Nuclear Test Ban

A worldwide nuclear test ban agreement is the simplest possible step of guaranteed arms limitation and would prevent the rise of other nuclear powers, or at least minimize their potential effectiveness.[1] It is the simplest step because it requires only a minimum deviation from conventional diplomatic and military attitudes, upon which our present partial security is based. It leaves us with our present nuclear weapons and the freedom to build more of them to keep the stalemate effective. It merely interferes with the rate of development of new weapons, treating the great powers equally so that neither can gain a decisive advantage. The step is simple also because it does not require the admission of inspectors with free access throughout the various countries.

The step is guaranteed against significant evasion because nuclear tests can be detected from afar. It is necessary to consider, at greater length than we shall here, the possibilities of special evasions, the limits of small air bursts that might not be detected by monitoring atmospheric radioactivity, the dependability of seismological detection of deep underground tests, and so forth. It seems very likely that a complete study would show that technically possible evasions would be of a minor nature and would not upset the stalemate. If it should, nevertheless, be deemed necessary, special provisions could be made to cover this difficulty which would only slightly complicate the otherwise simple scheme, such as admission of inspectors to seismic observatories at a few agreed spots in large countries. By and large, then, an adequate guarantee against evasion would be provided by an international monitoring agency whose primary tool would be airplane observation of atmospheric radioactivity and ground observation of fallout at various places throughout the world. Such an agency could very appropriately be set up under the United Nations. Its sole task would be to report any evasion, so that observing the ban would be in the self-interest of each nation.

The primary motive for the present H-bomb powers to enter into such a ban agreement would presumably be to prevent other powers from developing nuclear weapons. This motivation is primary because it is most immediate for the national interest. It is, perhaps, not the deepest motivation, nor the one that sounds best to the small countries. Yet, it is in the national interest of each small country to be kept from developing weapons so long as its immediate rivals are also prevented from doing so. No smaller nation can hope by its own nuclear arms to keep war from its borders or to maintain the world at peace. Its chances are better with our present arrangement in which two great powers keep each other in check. (England is a third nuclear power, so closely aligned with the United States as not greatly to increase the chances of trouble.)

A second major motive for a nuclear test ban is to try to preserve the technical possibility of more far-reaching agreement in the future. The nuclear test ban alone is not ideally suited for this purpose, but it would help a great deal. With this ban alone, development of ICBMs would continue apace. We cannot know just how well developed the warheads are that they would carry. The most authoritative hint that this development is still far from complete came recently from President Eisenhower. Replying on April 25 to Adlai Stevenson's inconsistent suggestion that we stop H-bomb tests but continue with ICBM development, the president said that "it would be futile for the United States to continue at 'top speed' on the ballistic missile if it were to stop testing the hydrogen bomb that would arm the projectile." He went on to say that "one weapon would be 'useless' without the other." It appears from these remarks that the ultimate destructiveness of the H-bomb has not been streamlined to fit into the severely limited payload of prospective multistage long-range missiles, and that a program of future nuclear tests is necessary to accomplish this development. Insofar as this is true, the nuclear test ban could indeed serve to prevent the transition to the next stage of technical difficulty in the control of weapons.

India has repeatedly suggested a nuclear test ban in the United Nations, and others have joined in the plea. The main motivation was to get rid of the dangers of radioactive fallout. These dangers seem very real indeed, particularly to the unfortunate Japanese. Yet the number of people likely to be severely affected by the continuing program of testing is so small compared to the expected loss of human life in nuclear war that the motivations concerned with avoidance of war remain the more powerful incentive to ban tests. That the unpleasantness associated with the tests themselves would also be eliminated is an added gain, but a small one.

Banning Missile Tests

A more effective way to prevent the modern Armageddon from growing beyond the technical possibility of control would be to ban the testing of long-range missiles. If instituted soon, this ban would effectively forestall their being brought into existence as usable weapons. The missile test ban has been vigorously proposed by Colonel Richard S. Leghorn before the Disarmament Subcommittee of the Senate Committee on Foreign Relations, March 7, 1956.[2] He pointed out that such a missile test ban could be monitored by a network of international radar picket posts set up at agreed points in national territories. This requires more relinquishment of national sovereignty in the traditional sense than does the nuclear test ban which can be monitored from outside. If this point looms large in the minds of statesmen, the missile ban could be considered as a separate and even secondary proposal. Yet the admission of international inspectors to relatively few agreed locations within a country is a very small concession compared to granting free access as would be required by more ambitious arms limitation plans. It seems unlikely that this concession to inspectors at isolated spots would loom nearly as important as deeper considerations of national safety.

Combined Nuclear and Missile Test Ban

The most convincing case can be made for the combined nuclear and missile test ban. Such an agreement between nations of the world would accomplish the objectives of both of the major motivations we have discussed, besides a host of minor ones presumably accompanying a reduction of tension engendered by the actual working of a very significant agreement. It would put a lid on the most threatening branches of weapon development. It would tend to leave long-range weapons technology about where it is today. Nations would be expected to work on development within the limitations of the agreement, but would find that they could not go far in these directions without further tests. More H-bombs could be built by the big powers, but not the most threatening vehicles to deliver them, for neither side yet knows how to make these missiles. Yet we could continue to rely on deterrence because we already have sufficient techniques. Save perhaps for quantity of bombers, we have essentially reached stalemate already. These considerations make it urgent that the combined test ban agreement be made an objective of national policy very soon.

Security with and without a Lid on Development

A short discussion like this of so weighty a matter, carried out without access to secret information, is always in danger of oversimplification. These are vital balances of advantage and disadvantage which need to be considered in detail and without bias. The fundamental question is whether we are safer with an agreement to limit the development of armaments or with the arms race wide open.

Weapons development, while largely engineering, involves science as its nursemaid and its source of new ideas, and science has a long tradition of unlimited development. It is heresy for a scientist to suggest that limits should be placed on a field of scientific investigation. Fundamental knowledge is good—understanding, an end in itself. One can never tell beforehand from what alley of idle curiosity the next fundamental advance will come.

It is tempting by simple analogy to extend these principles to weapons development. The analogy may be used to make a test ban sound morally reprehensible, like thought control or the burning of books. More practically, the danger is that confidence may be developed in control, only to have a new horror weapon spring up where least expected, perhaps suddenly to dominate the field. It must be admitted, for example, that our proposal puts no lid on the development of bacteriological warfare. If bacteriological developments are going to provide an ultimate weapon, they will provide it if we ban tests and do no more. If there is to be any deterrence it won't make much difference whether there has been a test ban or not. It seems likely that deterrence carried by bombers will be as effective as deterrence carried by ballistic missiles.

Senator Anderson has suggested[3] that research and development in new and demanding fields be promoted by establishing large international projects. He suggests weather control, controlled thermonuclear power, and space travel as examples. Many others, including bacteriological development and immunology, could be added. This plan would make a valuable addition to the combined nuclear and missile test ban. It would assure that all the cooperating nations would learn of any revolutionary technical developments at the same time. This would remove or greatly reduce one of the very real dangers of our present unrelenting national competition, the danger that one nation might suddenly find itself the sole possessor of a new principle with such important military applications as to make that nation feel immune from retaliation and tempted to aggression.

Unlike fundamental science, it is possible in weapons technology to delineate certain fields of development in which the important contributions in the fairly near future, at least, are expected to be of a preponderantly destructive nature. In banning nuclear bomb tests we might possibly prevent the discovery of a secret of nature essential to the development of controlled thermonuclear power. In any rational analysis it is well worth taking a small chance of such a loss if we thereby appreciably increase the likelihood that civilization may survive another century. If the missile test ban should demand it, we might likewise rationally decide to forgo interplanetary travel for a few more generations for the sake of keeping this beautiful planet habitable.

With the present unlimited arms race, our feeling of security lies in our confidence that we can run a good race. Through a combination of scientific and technical ability, devotion, the effectiveness of our free institutions, a head start, and good luck, we have had the experience of being definitely ahead. By trying our best, perhaps we can remain so indefinitely, but we cannot be sure. If an essential advance occurs first on our side, we feel that it is safe in our hands, for we will not attack. But one may occur on the other side and put us in danger. Even with development unlimited—from the point of view of cold technological competition alone (quite apart from mistakes of statesmen)—we are not entirely safe.

Our present aim is to remain ahead because we feel that it gives us an added margin of safety. It is not clear that there is as much short-term safety in remaining merely equal to our rivals in retaliatory power. As we approach saturation we inevitably approach effective equality, that is the symmetry of stalemate. The main difficulty is that we do not consider the two sides as equal in willingness to launch a surprise attack. If an attack might be launched against us, we feel that it is worthwhile for us to have twice as much force as necessary to destroy all targets, because the attackers may destroy half of our retaliatory force before we can strike back (or even more if their force is equal to ours). If we cannot keep far ahead in numbers of operational aircraft, we hope to do so in refinement of weaponry. Herein lies the partial security of technological dynamism that is the basis of our present policy. In this thinking we make the cautious assumption that for deterrence to be effective it is necessary for us to be able to inflict as much damage on them after being surprised as they can inflict on us in a surprise attack. This is effective deterrence.

This assumption is, however, not a part of the simple concept of mutual deterrence. The concept arises from the fact that nuclear weapons are so devastating that even a small part (say a quarter) of the destructive power of one of the rivals if delivered to the other would cause damage and human suffering so intolerable as to be unacceptable in any rational decision. It assumes that neither side can wipe out almost all of the other's retaliatory force in a simultaneous blow everywhere. Such deterrence is in principle effective between equals, even if one of them is looking for a chance to overwhelm the other. The only flaw is, as we have seen, that the congenitally aggressive side might achieve a technical breakthrough that would make it feel able to overwhelm the retaliatory force. Such a breakthrough would have to be a radical one, indeed, and in this case it is very doubtful whether having twice as powerful an established retaliatory force would be decisive in discouraging attack. There is thus an essential symmetry and rugged stability to the idea of mutual deterrence, a tolerance of slight inequalities, and it does not seem that much is added to our immediate security by our attempt to beat the symmetry and stay far ahead.

If we should decide to buy long-term national security by entering a test ban agreement, we would have to give up some of the competition in which we intend to stay ahead. Insofar as we are at present ahead, in entering a test ban agreement, we could expect to remain ahead for some years, but eventually to approach mere equality, and to be dependent on the deterrence of approximate equality.

If we continue with the arms race unlimited, we will probably maintain our lead in refinement of nuclear weapons (on the assumption that we have it) for a decade or so—as far as one can hope to foresee. Within this time the transition will probably be made to operational ICBMs, and we can feel no assurance that we will be the first to have them. This transition must be viewed as a danger period. Probably in the decade after that, and perhaps sooner, we should expect several other nuclear powers to enter the field. Then the future diplomacy of our present course begins to be really dangerous. This might engender strong pressures to control armaments, but it will be too late.

If, instead, we should succeed in negotiating a combined test ban agreement, we could presumably expect to continue with our lead in refinement of nuclear weapons and especially with our preponderant numbers, approaching equality toward the end of the decade. This approach to equality in itself would not constitute any real danger. The transition to intercontinental ballistic missiles would not take place, so the danger inherent in this transition would be avoided. The rise of other nuclear powers and the danger from a multilateral diplomacy getting out of hand would also be avoided. If internationalization of bacteriological and other developments is included in the agreement, the danger of unexpected surprises of this sort would be greatly reduced. If it should later become desirable and politically feasible to bring weapons under detailed international control, this would be technically possible with a practical amount of inspection because there would be no hidden intercontinental ballistic missiles to worry about.

The balance sheet is a complicated one to draw, and includes items hidden behind the veil of secrecy. However, this simplified look at today's balance book indicates far greater national security with the test ban than without—it would be hard to believe that the truth is otherwise.

NOTES

1. It is possible that a determined country could make a start in building an A-bomb arsenal without tests since the general principles of fission reaction are becoming fairly common knowledge, and it is known that past tests have succeeded. Development of the refinements, including H-bombs, is expected to be decisively retarded by lack of opprotunity to test new ideas, unless leakage of information should reveal great detail of design. The test ban would discourage the most modest program, and would so greatly handicap the new power in an attempt even remotely to approach the capabilities and confidence of the present powers as to practically preclude the attempt.

2. See article by Colonel Leghorn in *Bulletin of the Atomic Scientists* 12 (1956): 189–95.

3. See Senator Anderson's article in *Bulletin of the Atomic Scientists* 12 (1956): 223–26.

The Fourth-Country Problem: Let's Stop at Three

[1959]

Now that talk about stopping nuclear tests has spread from scientific circles to the forefront of international negotiation, a feeling seems to be arising that stopping tests has no positive value in itself, but is useful only as a stepping-stone to more effective disarmament arrangements. The United States is entering negotiations with the offer to continue a test moratorium beyond one year only on condition that, and as long as, "satisfactory progress is being made" toward further controls. France, widely but probably erroneously reported in the press to be on the very verge of its first atomic test, has stated that it will not abide by a mere agreement to stop tests unless this is a part of an agreement on more substantial disarmament steps.

Test Cessation a Vital Step

It is true that more than stopping tests—and much more—is required to provide real security for a world in which man is able to destroy himself. The need for introducing controls has an urgency arising from the fact that the task is not only very difficult but is becoming more difficult with the rapid advance and spread in weapons technology. It is desirable to do more than stop tests soon, but above all it is desirable to stop nuclear tests on a worldwide basis very soon. This step has a very real value in itself, quite apart from the fact that it will make further steps easier. It should be insistently pursued, as a minimum goal, to be sure, but independently, as a goal in itself regardless of what else may be done.

If no controls can be imposed on the arms race, we may expect continuing approach to perfection of instant destructive capability on the part of many nations, in spite of all attempts at developing defense against it. With vast and swift destructive force in the hands of many nations, annihilation of everything that makes life worthwhile is the almost certain end result. The deterrent effect of mutual terror may keep many nations out of trouble for a time, but sooner or later one of them will make the mistake fatal for all.

The development of ever swifter and more complete destruction could be slowed down but not stopped by a test ban. More extensive inspection facilities than needed for monitoring a test ban, delving deeper into national industrial affairs, would be needed to stop these developments. These would be much harder to negotiate, and agreement on them should be sought separately as a larger and perhaps lengthier undertaking. But the other serious increase of danger in the uncontrolled future is the ability that many nations will have to inflict large-scale destruction with nuclear weapons. This could be prevented by the prompt enactment of a well-controlled test ban. In this sense, the test ban, quite by itself for rather many years to come, solves an important part of the problem, and alone makes a valuable contribution to the chance that we may avoid nuclear world war. A test ban would influence the likelihood of war also in other ways, through its various effects on the relative military strength of the great rival nations, some of them influencing the balance in one direction and some in the other. There are the inequalities of capability of the two sides in various categories, and of the need for learning from nuclear tests, the question of the approach to saturation and the extent to which it leaves the inequalities important, etc. These have been and should be considered elsewhere and cannot be adequately reviewed in a few words here. But in the overall view, the danger of the many-nation capability to world stability looms so great, and the contribution of a test ban to this problem seems so important, as to make it very clear that a test ban would be in the national interest. . . .

In past negotiations we have several times made the mistake of demanding too big a package and ending with no agreement at all. The most glaring instance of this mistake was when Premier Bulganin, during a period of flux in the Soviet leadership, when success of outside agreements might have ameliorated internal developments, suggested ground control posts to warn against a surprise attack. Instead of accepting this and seeing how much we could make of it, we were so unprepared with any positive policy that we treated this suggested opening of the Iron Curtain as though it were greatly to the Soviet advantage, and would concede it to them only as a part of a package with aerial inspection.

As one studies the long and dismal history of disarmament negotiations throughout the last decade, one gets the discouraging impression that at almost every juncture neither side wanted progress toward agreement but merely propaganda advantage, and that each proposal was designed to sound good to the uninitiated but to be unacceptable to the other side.

As for the Soviet side, it is impossible for us to assess adequately their real policy. There is rapid vacillation in the tone of their approach, and there probably have been some changes within the Kremlin in the attitude toward negotiations in the period since Stalin's death, for better or worse. Many Westerners feel sure that any real international accommodation would be incompatible with the Communist expansive aims, yet this view must be tempered at least by the thought that they have no desire to take over a world in radioactive ashes. During periods of the belligerent Soviet approach to international talks it is easy to believe that there is little hope for any agreement. Yet we will never know what the possibilities are until we of the Western world have made as liberal a proposal as is possible, compatible with the aim of reducing the likelihood of war, a proposal with balanced incentives designed to appeal also to the Soviets' interest in avoiding real war.

On our side, our attempt to drive too hard a bargain has been a ruse for a lack of an agreed-upon policy objective. In some cases the officials promoting the negotiations wanted to bargain more flexibly, but could not obtain agreement at the cabinet level, due to conflicting departmental interests and attitudes. Perhaps our basic trouble is that we have no "Department of Nuclear Negotiation" to counterbalance the influence of the three powerful arms of the Department of Defense, with their traditional attitudes well suited to their historic functions. When the Stassen office started almost to function as such, though feebly with its inadequate scope and organization, it was promptly kicked downstairs as though out of fear that a positive nuclear negotiation policy would be embarrassing.

The Nth Country Incentives

The nightmare of a future with many nations capable of wreaking nuclear havoc and touching off the end of civilization has been haunting thinking men for several years. In 1954 the leaders of "Her Majesty's loyal opposition," the British Labour party, proposed that Britain should abstain from her first atomic tests on very sensible grounds: first, that Britain could thereby exert stronger influence to bring the two nuclear powers to agreement on arms limitation, and second, and perhaps more important, that if Britain should be the third to join the nuclear club, there would soon be a fourth to insist on the same right, and so on. Since the start of the British tests, the problem of forestalling the nightmare has been known widely as the "fourth-country problem," but in France it has been known as the "fifth-country problem." If France does carry through and become number four, as widely anticipated, there will probably be no stopping Communist China, perhaps as the fifth country, and some of France's closer neighbors and rivals not far down the line.

The Nth-nation problem thus promises to be practically an infinite series leading to a catastrophic divergence unless we succeed in cutting it off soon.

We didn't wake up in time to cut it off at three. If we do not succeed in cutting it off at four, when the country concerned is France, still vaguely within our "sphere of influence," it seems likely that we will not succeed in cutting it off at all. . . .

But why should the French permit themselves to be convinced to desist? They should do so because the same incentive that applies to the nuclear "haves" should also motivate the "have-nots." The real interest of all nations, if they but know it, lies in minimizing the likelihood of a nuclear world war. Each nation must think not just what it loses by not carrying out its own tests, but also how much more it gains if others do not carry out tests. For France, the argument comes on two levels, each of them persuasive. First, there is the reduced chance that the United States and the USSR will get into a war with each other if, by agreement on a test ban, the tense world situation is kept from getting worse. Second, there is the thought that France's action in developing nuclear arms will inevitably lead to a similar acquisition a few years later by some of her neighbors such as Italy, Switzerland, Germany (East and West), Poland, Sweden, and eventually Spain. Then any temporary gain in security or prestige will be annulled by the increased danger from local rivalries.

There is a contrary argument that cannot but weigh heavily on the minds of the "have-nots." It is based on the fear that one cannot trust the will of an ally such as the United States to defend one's own nation by steadfast threat of retaliation if the stakes become as high as they are in nuclear war. This argument is weakened, however, by the doubt that a small "nuclear" nation would actually defend itself against a large one, and risk bringing annihilation on itself. In the long run, it seems, the first argument should prevail; it is safer to trust a well-defined system of mutual deterrence between the present nuclear "haves" to maintain the balance and shield the "have-nots" than to expect many nations to shield themselves, with the concomitant danger that any pair of them might set off a fast chain reaction to engulf all. This choice is the more attractive because the test ban that would be involved as its first step holds the promise of leading to further steps toward arms control and the eventual organization of sound arrangements for world stability.

Thus the rational choice for France should be not to insist on atomic tests. Yet there are great nationalistic and psychological barriers in the way of making the rational choice, even as there have been for us, and perhaps greater. The fortunes of history, more good than bad, have left France in the habit of being a proud nation. Times have suddenly changed, war is no longer a practical instrument of foreign policy, and nuclear arms are not for use against Algerians. However, the possession of nuclear arms has become a symbol of being a great nation, influential at the council table. With this symbol almost within her grasp, France will find it difficult to withdraw her claim. It will take a great act of leadership to convince France that her future will be more secure if she lets her greatness rest in her art and culture and commerce.

For each nation of the world the decision is difficult because it demands departure from traditional attitudes. It is the civilized world as a whole that is in unprecedented danger as a result of unprecedented technical developments. Each nation must understand that its security depends on world stability. Each must realize, in judging the balance of its own security, that it may profitably give up some element of its own traditional military strength if others likewise give it up. There has been too little willingness, for example, to renounce nuclear testing except just at the time when the completion of a test series has temporarily fulfilled a given nation's need for tests. Nations must come to the council table willing to renounce something desirable in exchange for a similar renunciation by others, if negotiations are to succeed. May the nations large and small, whether nuclear or nonnuclear, rise to this necessity.

Nuclear Threats, ABM Systems, and Proliferation

[1968]

The destructive power of nuclear weapons is so terrifying that the most essential objective of our national nuclear policy is to avoid their use in all-out war. The purpose of our vast strategic striking force is to deter strategic nuclear attack against our cities. Our smaller but still very powerful tactical nuclear weapons were stationed in Europe to deter the Soviet Union from launching a land attack there. This is the great "nuclear umbrella" intended to prevent the outbreak of nuclear war.

Those who, in the early days of the atomic age, thought in terms of "one world or none" contended that with so terrible a nuclear threat all wars must cease, for a conventional war would too easily escalate into nuclear war. It was argued that the hard-pressed side, rather than admit defeat, would be sorely tempted to resort to nuclear weapons. Instead, the prevailing doctrine has been that the competitive pursuit of national interests in the real world will continue to lead to small conventional wars—"brushfire wars" was the popular term—under the cover of the nuclear umbrella that should prevent their becoming nuclear.

We are now engaged in an oversized "brushfire war" in Vietnam, a war that is alarming both in the extent of misery inflicted and the amount of national strength deflected from better ends on both sides. In it the intensity of the action has had its ups and downs. As this is written the exposed bastion at Khe Sanh is no longer a hot spot, but a few months ago it was looked upon as a potential Dien Bien Phu, and the president took the unusual step of exacting from each of the members of the Joint Chiefs of Staff an assurance that it would not fall. In the tenseness of that situation—and a similar situation may arise again—there seemed to be genuine uncertainty as to whether we would use tactical nuclear weapons against the Vietnamese enemy in a pinch. The matter was discussed, by Senators Fulbright and McCarthy, among others, and when pressed the president would only say, negatively, that the use of nuclear weapons in Vietnam was not under consideration and he had had no requests for their use there. At about the same time General Wheeler said that he did not believe that nuclear weapons would be necessary for the defense of Khe Sanh. While it is gratifying that he turned out to be right and that nuclear weapons have not been used in Vietnam, his statement, coupled with the president's faint denials, reflected a fluid state of policy. Less prominent generals made more belligerent statements, while background briefings by U.S. spokesmen suggested that tactical nuclear weapons would lose their deterrent effect if decisively renounced and the secretary of state objected to having so sensitive a matter discussed in public.

Our leaders know full well the importance of keeping inviolate that precious line of demarcation between conventional weapons and nuclear weapons, clearly marked as it is by the radioactivity of even the least powerful nuclear bursts. They know that escalation must stop at that line. They know that our homeland is vulnerable only to nuclear weapons, and that it would be unutterably stupid for us to initiate their use in this age of nuclear competition. They know of the risk of bringing China and Russia into the war, and they know, too, that if we use tactical nuclear weapons they could be supplied to the other side, and that our great concentrations of supplies and troops would be ideal targets. Yet our leaders' faint denials inspire no confidence that in the anguish and heat of battle they would remember what they know.

But even if we assume that the administration will not be so irresponsible and foolish as to use tactical nuclear weapons against Vietnamese, it is still a great mistake to leave the question open. The president's faint denial and the more direct suggestions by subordinates of possible nuclear strikes had the effect of being a slightly veiled nuclear threat against the Vietnamese enemy, probably an intentional one meant to "keep them guessing." This introduces an entirely new dimension into the strategy of the nuclear age, for it is employing a nuclear threat, not to deter attack by a nuclear nation and prevent the outbreak of a war, but to deter a nonnuclear country from some act of conventional local attack within a "brushfire" battle already in progress.

In merely implying this threat, whether intended or not, the administration is still being extremely shortsighted, irresponsible, and foolish. Very high on our list of national priorities, and properly so, is avoiding the proliferation of nuclear weapons to many countries. A future with many nuclear nations would be a diplomatic nightmare; in it the triggering of a worldwide nuclear war would be all too likely. The nonproliferation treaty negotiated at Geneva is an important beginning if enough nonnuclear nations can be induced to sign and to stay with it.

Whether they will sign the treaty and long submit to its restrictions will depend upon the importance that is attached to possession of nuclear weapons. As long as nuclear weapons are just a part of the nuclear umbrella—a mutual threat by nuclear powers to prevent the outbreak of war among them—

the nonnuclear nations may well feel glad to have no part of it. But now the foolish equivocal stance of our administration has introduced a nuclear threat against a nonnuclear country into conventional warfare. For other countries this puts a premium on having nuclear weapons with which to face such a threat. In debating their adherence to the treaty, some of the most important nonnuclear nations have been calling for some step of nuclear arms restraint by the nuclear nations to match the restraint demanded of them. At the very least we should show enough restraint to clear up this matter and retract an implied nuclear threat against a small nonnuclear country. . . .

By the time this issue of the *Bulletin* appears, the General Assembly debate of the Nonproliferation Treaty should be completed unless the few nations that object have obtained a postponement.

The declaration of intent, Article VI of the treaty draft, reads "Each of the Parties of the Treaty undertakes to pursue negotiations in good faith on effective measures relating to cessation of the nuclear arms race at an early date and to nuclear disarmament, and on a Treaty on general and complete disarmament under strict and effective international control."

For including only this declaration, rather than, say, an obligation to forgo ABMs and stop nuclear weapon production, the superpower argument has been that nonproliferation is urgent and it would take too long to fill in agreed details of such an obligation, so that the other nations must take the intention on faith until that "early date." In view of the sad record of superpower attempts at arms limitation and the recent plunge into the ABM-MIRV phase of the arms race, some lack of faith is understandable.

The intent so declared to seek cessation of the nuclear arms race is doubtless very sincerely held in those agencies of the superpower governments that have been most active in promoting the Nonproliferation Treaty, and most appreciative of the dangers of a future world of many nuclear powers. The question on which the future viability of the treaty hangs is whether these agencies can convince their governments to go along with some substantial measure of restraint for the sake of avoiding these dangers.

In this connection it is slightly encouraging to note that the votes in the Senate on amendments opposing the ABM deployment were considerably closer on April 18 of this year than a year and two years earlier. These were amendments to the military "authorization" bill and only preliminary to a later vote on an appropriation bill. The closeness of the votes encourages renewed efforts to block the ABM funding in the appropriation bill. It is a pity that a majority of senators still seem not to appreciate that our real problem now is not our relative strength vis-à-vis the USSR or China, which is more than adequate; not the stimulation of industry through defense spending; but far more importantly, the problem is the worldwide spread of nuclear weapons that we cannot stop without stopping the arms race by first stopping the ABM deployment.

4 ABMs—Antiballistic Missiles

Ever since the era of the shield against the sword, counter-measures have been developed to blunt the effectiveness of military weapons systems. Antiaircraft batteries in World War II could be counted as successful if they could bring down 10 percent of an attacking bomber force because chemical explosives inaccurately delivered amidst confusion did not cause enough damage to justify that great a loss by the attacker. A similar defense against a missile attack with H-bombs would have to be close to 100 percent effective to prevent unprecedented destruction. However, there was an urge to try. The fact that the Russians, with their defensive psychology imbued by their sad history, installed a presumably ineffective antiballistic missile system around Moscow provided ample promotional grounds for our having more of the same. An elaborate system was devised involving both large Spartan missiles carrying H-bombs to intercept incoming missiles in space hundreds of miles from their targets and short-range Sprint missiles, also carrying H-bombs, for interception in the atmosphere near the target. Even though penetration aids—countermeasures to the countermeasures—made an attempt at ABM defense seem technically futile, Congress was nevertheless impressed by Pentagon demands.

A politically devoted poet named Lenore Marshall, whom I had the pleasure of knowing on the board of directors of the National Committee for a Sane Nuclear Policy (later known simply as SANE), wrote on this subject of countermeasures:

> I was happy, I could whistle
> Until he made his anti-missile.
> I felt better when I read
> Anti-antis were ahead.
> Now I'm safe again, but can't he
> Make an anti-anti-anti?

Indeed, the elaboration of countermeasures to countermeasures in the ABM technology aptly illustrated how one development continually leads to another in the arms race.

But the greater objection to ABMs was that, unless they were obviously completely ineffective or practically 100 percent effective, they would destabilize the deterrent and accelerate the arms race. The very uncertainty whether they would work, coupled with worst-case analysis, would magnify demands for offensive capability/Leaders on each side would conservatively assume the opponent's ABMs to be more effective than their own and would want to have more strategic missiles than the other side, so there could be no balance. This influence that makes it difficult to stop the arms race with a stable balance is sometimes called arms race instability. ABM systems would also tend to make a first strike more tempting in a crisis, for penetration of the defense would be more certain by a massive first strike than by a depleted second strike. This is known as crisis instability. ABMs increase both kinds of instability/

Deterrence is usually envisioned as a symmetrical balance of terror between two adversaries, each having the assured capability of doing unacceptable damage to the cities of the other so each is deterred from starting a nuclear war. Actually, this is a bit oversimplified. The deterrence that has been in effect for a quarter of a century is not so symmetrical because the two sides profess different strategic doctrines. Our policy has been primarily to deter nuclear attack by threatening Soviet cities. Soviet policy, being deeply influenced by a national memory of tragic past invasions, seeks mainly to minimize damage to their cities, if attack seems inevitable, by making a preëmptive first strike at our missiles. Deterrence is then an asymmetrical balance. We are deterred from letting a crisis develop by fear of triggering a Soviet first strike which would incidentally slaughter our people, and they are deterred by the knowledge that their counterforce first strike can do no more than somewhat reduce the damage by our second strike. There are also further complications, but the tendency remains for ABMs to increase both kinds of instability.

The first of the following articles discusses the situation in 1967, when momentum for ABM development was beginning to build up despite such arguments as those presented in that article. It seemed again as though we objectors were no match for the promoters. But then a lucky thing happened in our group of concerned scientists at the Argonne National Laboratory near Chicago. John Erskine read in the local west suburban paper that some army engineers were investigating a nearby site for installing missiles. He stopped by one day to talk to the colonel in charge of the test drilling and was astounded to learn that the intention was to install Spartan missiles there, with their H-bomb warheads, not just the old Nike-Zeus missiles with chemical explosives such as had already been installed near cities. We discussed the situation at our little luncheon group of Federation of American Scientist members and called a press conference at John's home where television cameras, as well as reporters, appeared. The news hit Chicago impressively and was picked up nationally. There was a flood of calls for us to speak at various local groups and John Erskine and George Stanford responded particularly generously, sometimes debating with representatives of the Department of Defense. At that time there was also agitation, led by Senator Dodd, for a neutron bomb. On one occasion in the question period after a talk, George was asked, "What is this neutron bomb they talk about good for anyway?" He quickly responded, "Dodd only knows."

The second of the following articles appeared as part of that effort. Its title had perhaps more influence than anything else in these articles. The slogan "H-bombs in the back yard" echoed elsewhere in the national press as the campaign against serious deployment of an ABM system gathered momentum nationally and was finally successful. Some influential insiders, notably the President's Science Advisory Committee (PSAC) were valiantly opposing ABMs in government councils, but the outside citizens' pressure turned the tide.

Of the many citizens' initiatives to introduce some measure of restraint into the arms race, the two that have been most successful have been those concerning matters that come close to home for the average citizen. The partial test ban resulted from citizen pressure to avoid fallout close to home. Negotiation of the agreement in SALT I not to build elaborate ABM systems (beyond one local installation each in the United States and Russia) was influenced by citizen pressure to avoid "H-bombs in the back yard." Many citizens of many countries are now becoming increasingly aware that the threat of a perpetual arms race comes close to home. The present-day grass roots pressure for a freeze on further deployment of nuclear weapons systems is being heard by some political leaders, unfortunately still too few, and some day may likewise succeed in turning the tide.

Now in the mid-eighties the ABM Follies, this time billed as "Star Wars," or the "Strategic Defense Initiative," are being restaged in modern dress featuring far more spectacular extravaganzas. The earlier ABM script back in the sixties was based mainly on current technology. The objections that made it a flop were that even an enormous effort could make it only partially effective and that it would destabilize the arms race. The "Star Wars" dream is based on notions of future technology spun of pure fantasy. Its proponents find it politically attractive to imagine a defensive shield so perfect as to make nuclear weapons obsolete, a fantasy that a confused public is understandably anxious to believe. Their more tangible incentive to pursue it is the economic importance of keeping open a pipeline of dollars from government to industry and to weapons labs that have outlived their military usefulness. The prospect of ever attaining the distant goals while continuing with the arms race is not to be taken seriously, but the destabilizing impact of early steps in the Strategic Defense Initiative, particularly the increased vulnerability of satellites, is very serious indeed.

The one situation in which an ABM system would be useful to help preserve the peace would be in a nearly disarmed world. Then a threatened nuclear strike would consist of so few missiles that an extensive ABM system could be expected to intercept all or almost all of them. Back in the fifties and sixties questions of how to approach a disarmed world were much discussed by those concerned about the arms race. Some of the thoughts of those days appear in chapter 5. The last stage of any disarmament process was recognized to be the most difficult, when each nation would be expected to give up its last nuclear weapons without being quite sure that another country might not cheat by secretly retaining or making a few and thereby possess overwhelming power. The emphasis then was on tackling the easier challenge first, on devising ways to get down to a minimum deterrent, a level of armament just sufficient to inflict clearly unacceptable damage and to cover the uncertainty about cheating. It might involve ten missiles on each side, or fifty or a hundred but not thousands. It was sometimes called a transitional deterrent to express the hope that it might be a way station on the way to complete disarmament, a level at which to pause, perhaps for quite a long time, while developing the confidence and the ideas needed to take the last step. In those early discussions ABMs were not yet in the picture. They would be useful in taking that last step to complete disarmament. Against a threatened attack by the fewer missiles of a minimum deterrent they could be effective. Introducing them would accomplish a transition from deterrence by threat of retaliation to deterrence by impenetrable shields. During the transition there would be a crossover period of uncertainty, so the transition should be arranged to be as quick and orderly as possible. The pause at a minimum-deterrent level should allow time to arrange all that, perhaps even by international development of the technologies still to be used. Until then we should negotiate a freeze on such developments, including "Star Wars," as the

Soviets have indeed proposed. After the transition, the elimination of remaining strategic missiles should be much simpler since, under the ABM shield, a few clandestine missiles need not be feared. A concurrent problem would be how to avoid other means of delivery. If we now permit cruise missiles to proliferate, that may never be possible. They will make obsolete not only "Star Wars," for they avoid attacking through outer space, but also the hope of verifiable arms control.

Now, in 1985, the prospect looks bleak for any early termination of the arms race. The administration seems bent on pursuit of its fantastic "Star Wars" initiative, extending the arms race into vast new realms. This is so clearly bad for national security that the main reason for pursuing it—aside from the political appeal of false perceptions—must be that it absorbs money and specialized effort, in an attempt to stimulate the economy. Appropriations for cruise missiles are taken for granted in the swollen budget. If the political realities indeed demand such military expenditures, the administration would do well to change its course and direct the money and effort instead into channels that would be politically perceived as national defense and would be less damaging to future options. Despite the serious drawbacks of ABMs in general, one possibility would be to spend a lot of money and technical effort on ABM defense of missile sites, while reducing strategic missiles. This scenario is not nearly as good as some of us have long been seeking, but not as bad as the way we are going.

Such a scheme has been proposed by J. N. Barkenbus and A. M. Weinberg under the name "defense-protected build-down."[1] If we and the Soviets have about equal capabilities of knocking out each other's missiles it seems important to maintain a perceived balance. Unilateral reduction of our missile force is politically unacceptable because it would seem to disturb the balance. It would weaken our side. For us instead to introduce missile defense unilaterally would disturb the balance in the opposite way. In effect it would weaken the other side, for fewer of their missiles would get through. The proposal is to do a little of each simultaneously, in a combination estimated to preserve the balance. If we should reduce our strategic missile force by 10 percent and install ABMs capable of intercepting about 10 percent of attacking missiles, then both attack forces would be reduced in effect by about 10 percent and would still be balanced. The Soviets might object to our ABMs but would have a sweetener in the reduction of the missiles aimed at them. They might even like the idea enough to follow suit, thus initiating a sequence of reciprocated unilateral reductions that might lead to a negotiated continuation of the process.

The influence of the defense component on the politics of vested interests may be even more important. The military-industrial complex—and indeed much of the public—would dislike ending the arms race and just reducing our attack force unilaterally, but would see a sweetener in the opportunity to find prestige, profits, and a perception of security in developing and deploying the defense. The defense proposed would be ground-based terminal interception and should not be allowed to provide an incentive for the Star Wars type.

It has been said that a partially effective defense would contribute both arms-race instability and crisis instability. The defense-protected build-down eliminates the former but not the latter. However, this crisis concern is less important than ending the arms race. It is a concern for a relatively short transition period during which a severe crisis is not very apt to arise, whereas the arms race must end for the sake of the future.

In the past forty years there have been many proposals, including some in the articles in this book, of ways to end the arms race and start moving toward a safer world. They have failed to halt the momentum of the arms race not because they could not work but because in high places there has been no shared will to disarm. Doubts about verification and disagreements over assessments could be used against them. This latest proposal has these same obstacles to overcome, but the sweetner in the costly deployment of defense might spell the difference this time. It might combine with growing military doubts about the practicality of various war-fighting strategies and lead to the needed will to disarm.

Since ABM systems facing a large missile force introduce uncertainty which is destabilizing, it is important that they should be developed and introduced only in the right way and at the right time, if at all. Unless using them in such a politically expedient manner is the only way the arms race can be ended soon, they should not be developed until missile deployments have been reduced to such a low level that an ABM system could be effective. The value of extensive ABM systems in a disarmed or nearly disarmed world is recognized in recent interesting books by Jonathan Schell[2] and Freeman Dyson.[3] Neither of them sufficiently divorces the utility of ABMs in a hypothetical future from the danger of their development in the real present.

Dyson makes much of the prospect that antimissile missiles might in the future no longer need to carry nuclear explosives or indeed any explosive at all, making up in accuracy for their lack of explosive power. He points out that if an inert projectile of only a few pounds could be placed directly in the path of an incoming missile so accurately as to hit it on the nose, the missile would shatter itself on impact because of its great speed. Such a development, if possible, would mean lighter payloads and smaller missiles so a defender could perhaps afford more of them despite the cost of the advanced associated equipment, making a more effective shield.

Schell emphasizes that even in a world without nuclear explosives there would still be nuclear deterrence, deterrence based not on the threat of immediate attack but on the threat of rearming and then attacking. The facilities that could produce nuclear weapons are what pose the threat, and Schell foresees a need for international control of these facilities replacing the present need for control of the weapons themselves. The lead time for attack would thus be increased from minutes to months, eliminating hotheaded attack and leaving some time for reason to prevail in a crisis. Nonnuclear ABM systems could be an important part of that ideally disarmed world. They would help stabilize this "disarmed deterrence," for it would take a long time for a plotting aggressor to accumulate a massive enough striking force to penetrate them.

Schell has no use for what he calls a minimum deterrent, but he defines it differently. He uses the term to mean minimum annihilation, a striking force able to destroy an adversary only once, without multiple overkill. Our smaller minimum deterrent would be a useful step toward his goal, providing a time for developing the ABM system that should not be developed earlier. Even if the last step to complete disarmament is never taken, we would be safer with a permanent minimum deterrent force stabilized by an effective ABM system than with the present arms race.

Assessing the effectiveness of the more sophisticated and complex future ABM systems now being proposed involves some of the same fundamental considerations that stood in the way of earlier ABM proposals, particularly the impossibility of developing confidence in high enough performance of very complex systems that could never be realistically tested and the destabilizing dangers of the uncertainty they introduce. For the sake of perspective in facing our present problems, it is thus good to review the thoughts of an earlier era presented in these articles and to appreciate that we are better off now, in terms of stability, for having banned ABMs then.

NOTES

1. Jack N. Barkenbus and Alvin M. Weinberg, *Bulletin of the Atomic Scientists* 40 (October 1984): 18.
2. Jonathan Schell, *The Abolition* (New York: Knopf, 1984).
3. Freeman Dyson, *Weapons and Hope* (New York: Harper and Row, 1984).

Antimissile Drag Race

[1967]

The two nuclear giants face each other, glowering, but each with a healthy respect for the striking power of the other that adds up to relative stability of the nuclear deterrent. They are both worried about the birth in the near future of a swarm of nuclear dwarfs that will someday grow up and upset the stability, and are wondering how to prevent it. To induce the nuclear dwarfs—the Nth countries—to desist from entering the nuclear race, the giants must somehow show some restraint and cut off the race between themselves. They have no real need to race if they can agree where to stop—either tacitly or by treaty. The power of modern weapons is so tremendous that each side has more than enough to induce a healthy respect in the other, and it is not important for stability that the two sides have equal strength, nor that our side have much more than the other side.

Actually, the United States is said to have something like three or four times as many long-range missiles as has the USSR—a more than comfortable margin from our point of view. This came about more by accident than design, as a result of our overresponse to the imagined missile gap shortly after the Russians impressed us with their Sputnik. It leaves us sitting pretty and with a habit of considering it natural that we should have an enormous margin and that the Russians should accept this. It is very doubtful that they will accept so large a margin in the long run, and not necessary or reasonable that we should expect them to, but we can make it a race to preserve the large margin if we insist. We seem to be on the verge of doing this and forgetting more vital aspects of nuclear stability.

Until recently we seemed to be approaching a steady level on both sides, but now this situation of contentment is in danger of being upset by a new element: antimissile defense. Strategic ballistic missiles with hydrogen warheads are so immensely powerful, and so difficult to intercept when they come in large numbers, that it may be said now, and probably for a long time, that there is no effective defense against them. Efforts are in progress to develop an effective defense, and if they should succeed they would lead to quite another equation for nuclear stability; but the chance for success is not very great. In the meantime, there is a partially effective system that is capable of modifying the balance somewhat. This is our Nike-X system and its Soviet counterpart. If massively deployed, it is expected that it will be able to intercept, by explosion of its own hydrogen warheads in the atmosphere, some modest percentage of the missiles of a massive attack. Attack missiles are being designed with penetration aids to keep the percentage small. We are already introducing the first of these in our Poseidon missile to reduce the effectiveness of a Russian antimissile defense.

If one side deploys an antimissile defense, the other side would be able to deliver somewhat fewer missiles on target but could overcome this disadvantage by mounting more attack missiles. Antimissile defenses now in prospect are so inefficient that the cost of mounting the additional attack missiles to redress the balance is estimated to be much less than the cost of the defense— which, in the case of the United States, would amount to some $40 billion in ten years.

Under these circumstances, it may seem strange that the Russians should be so spendthrift as to mount an antimissile defense. But it is reliably reported that they are starting to do so, doubtless on a scale much smaller than they will ultimately reach if we insist on making a race of it. Although antimissile defense is militarily uneconomical, for them it has a special psychological value. Because of their traumatic experiences with the invasions of Napoleon and Hitler, they have a compulsion to buy anything that has the label "defensive."

Antimissile defense is, indeed, purely defensive. The argument for it is that, *if* an attack of a certain type is made, the antimissile defense and its associated shelters will save lives. But that is a big *if*. The more important question is whether the antimissile defense will make the attack less likely or more likely. The preponderant argument is that, by continuing and destabilizing the arms race, the antimissile systems will make nuclear war more likely. It is natural to expect that competitive deployment of antimissile defense would eventually be accompanied by a step-up in strategic missiles to penetrate the defenses and by a tremendous shelter program which implies undesirable population regimentation. Meanwhile, political obstacles to constructive diplomacy are probably increasing. These considerations and the crying needs of a dangerously and increasingly hungry world are all good reasons why we should not start down the rosy path of antimissile defense.

In his state of the union message, President Johnson referred to Russia's deployment of "a limited antimissile defense" around Moscow and said that a new round of the arms race based on this would be "an additional waste of resources with no gain in security to either side." He has the approval of most senators in his intention to seek an agreement with the USSR to ban further antimissile deployment, but there also is an attitude in Washington that we cannot possibly do without any bit of military hardware that the Russians have, even if they don't have very much of it, and we think they are foolish and wasteful to have it. The U.S. Joint Chiefs of Staff seem to want any military hardware they can get, and have recommended antimissile deployment. Secretary of Defense McNamara, being interested in the efficiency of the military establishment, has been holding out against it, with the backing of the president.

Influenced by the Joint Chiefs (and probably by other considerations, including local economics), Congress last summer tried to jam it down the throat of the administration by passing a supplemental appropriation, not requested by the president, for some initial stages of deployment. Only fourteen thoughtful senators voted against it. The administration still hasn't spent it, but shows signs of feeling the pressure, and much larger appropriations are being discussed for this session. The Senate seems to look on the antimissile race as a drag race, being concerned mainly with getting started and not with where we are going in foreign policy. But once the momentum of hardware procurement is built up, there will be no stopping and very little freedom to steer.

National politics being what it is, an anti-antimissile agreement with the Russians probably is the only way that this next upward spiral in the arms race can be avoided, and it is fervently to be hoped that they will agree. Yet it seems rather unlikely that they will, in the light of their complex psychological and political problems involving Napoleon and Peking, not to mention Vietnam. If this be the course of events, as seems likely, it will be a pity that the foreign policy of our great nation should miss urgent opportunities for constructive diplomacy in a world of want by being so paranoid and imitative.

The alternative would be for us this once to forgo a compulsive response, to permit Moscow the psychological lift of girding itself with an inefficient antimissile system, and to watch the deployment stop at modest proportions for lack of competition. Our more-than-comfortable margin of superiority in strategic missiles permits us the option of waiting to observe this development without losing our position of strength. In the politics of tacit agreement and arms control by example, we could make it clear that our lack of response is predicated on their stopping within reason. Because of their peculiar bias toward defensive measures, they could be content with a static situation in which our side would retain an overall margin of superiority, and both sides should be happy. Then, in an atmosphere of relative cordiality, we and Russia could pursue the difficult and delicate problems posed by our truly mutual interests in avoiding the proliferation of nuclear weapons and in otherwise calming a hungry world. This restraint on the part of the giants may be used as a basis for wheedling restraint from the nuclear dwarfs.

H-Bombs in the Back Yard

[1968]

A reporter for a small suburban newspaper recently visited a drilling rig on the edge of Clarendon Hills, a western suburb of Chicago, and inquired what was up. He learned that the army was exploring for a suitable site for antiballistic missiles. A scientist from the suburban Argonne National Laboratory noticed the story; subsequent luncheon-table discussions aroused concern among scientists which soon spread to the Chicago news media. Insidious are the ways of military public relations, and this is how Chicago happened to learn that, if all goes as planned, it is to have H-bomb-tipped missles installed in its own back yard, on the edge of Cook County upwind from the Loop.

When one of the scientists went to talk to the colonel in charge of the drilling operation, he was astounded to learn that the Sentinel installation was to include long-range Spartan missiles, in addition to the short-range Sprints. Only the Sprints might conceivably have some reason to be near a city if Congress should in the future opt for an attempt at city defense and authorize something much larger than the $5 billion Sentinel system. Later word from Lieut. Gen. A. D. Starbird, after a secret briefing in Chicago on November 29, is still more surprising: The site will have only long-range Spartans, no Sprints. Some other sites may get Sprints.

The capability claimed for the Sentinel system is that its Spartan missiles can stop a small attack by a few missiles—such as the Chinese might have in the mid-seventies—if they are as primitive as our first ICBMs in lacking penetration aids. The Sprints of the system are mainly to protect its Spartans and the accompanying radar. An optional "kicker" in the system, as was explained by its promoters, is that its short-range Sprints might be used to provide some protection for our ICBMs in their underground silos, and thus slightly blunt a Soviet counterforce attack.

There are, of course, far-reaching implications of the decision to deploy an ABM system, implications for the stability of the nuclear deterrent, for the future of the entire arms race, and for the possibility of diplomatic initiatives that might reduce the likelihood of nuclear war. But there are, in addition, two purely local objections. First is the possibility that, in a limited nuclear war with the Soviet Union, local Spartans might draw enemy fire to the city. The army's reply is that the population centers are prime targets in any event. But who knows? There has been long and vacillating argument about the "counterforce" and "counterpopulation" options of nuclear attack. Should an attacker spend his first salvos on the missiles of the enemy in an attempt to minimize retribution, or should he concentrate on doing "unacceptable damage" to the population and expect to take the brunt of a counterattack on his own population?

The think tank pendulum has swung between one and the other. Counter-population is the current style on our side, and that is what the army means by saying the cities are prime targets in any event. But, who knows, the Soviet high command might believe in counterforce ten years from now. If they should follow this course, and on some tense occasion attack, they might decide to strike at the Spartans on the edge of Cook County that could conceivably defend some of our ICBMs. In the process they would devastate Chicago and pulverize some western suburbs. If, on the other hand, they decide to attack both types of targets, we will have helped them kill two birds with one stone.

An objection based on the distinction between limited and all-out nuclear war may seem not very serious because any nuclear war would represent a disastrous failure of policy, and it is hard to believe that it could remain limited. However, there is also no serious reason for the Spartans to be close to cities, since their effectiveness must be nearly uniform over the central part of the 600 to 1,000 mile-wide range they attempt to defend. This is implied in various official statements and in information given to Congress during debates leading up to the initial appropriations for the system. There was very little discussion of where the sites would be, but Congressman Sikes, floor leader for the Sentinel appropriation, stated in the House on July 29, 1968, that "these sites will be some distance away from the centers of population."

In reply to the sudden publicity, the Chicago *Sun-Times* of November 16 quoted Col. R. J. Bennett, information officer of the Huntsville, Alabama, missile center, as saying: "The Sentinel site near Chicago is necessary to complete the Sentinel defense of the entire United States. To make such a defense most effective, considering the projection of future defense needs, this site should be near the center of the greatest population."

Here is the tip-off of the army's intentions. Congress has authorized the deployment of the Sentinel system and has funded its initial stages, particularly site acquisition. In the Senate debates, the main motivations for deployment given by the promoters of the system were defense against a Chinese attack and the protection it might afford against an accidental launching of a Soviet ICBM. There were a few senators who frankly argued for it as a step toward a much larger anti-Soviet system, which is probably the real reason the inherently expansive Department of Defense supports it. The initial Sentinel, it was said, might serve as a "building block" for the much larger system. Still, it seems clear that most of the senators who voted for the deployment—and the votes were fairly close—did so out of a feeling that, being in doubt, they should now support only the limited Sentinel system and either oppose the larger system or put off the larger decision. Thus in using a "projection of future defense needs" to justify putting Spartans near large populations, the army seems to be jumping the gun on a congressional decision and acquiring sites for the larger anti-Soviet system, under the guise of limited Sentinel deployment.

A second local objection to these sites is that there is some chance, probably very small, that one of the cluster of H-bomb warheads installed on the edge of the city might accidentally explode, and if it should, the consequent loss of life could be catastrophic. A surface burst or a shallow subsurface burst in a silo produces much more fallout—from vaporized and activated earth—than a normal explosion high in the air. The Spartan warhead is said to be "in the megaton range." This would indicate a weapon approximately a hundred times as powerful as the bomb that destroyed Hiroshima from half a mile in the air. Its local fallout from an accidental subsurface burst would be highly lethal throughout a large metropolitan area and for many miles downwind. There would be less blast damage than from an air burst, but it would still be widespread enough to flatten several suburbs.

An accidental explosion of a Sprint would, of course, be much less lethal. How much less is hard to say because we are told only that its warhead is much less powerful than a Spartan—"in the kiloton range." Taken literally, this could mean anywhere from one kiloton, or perhaps even less, to a hundred kilotons or more. Indications are, however, that it is considerably less powerful than the twenty kilotons of the Hiroshima bomb or the first A-bomb tested one hundred feet above the New Mexico desert. Even so, it could pose a serious hazard in the vicinity because of the high amount of fallout produced by a shallow subsurface detonation. Whatever the uncertain magnitude of this Sprint hazard may be, an accidental burst of the monstrously powerful Spartan warhead would be calamitous indeed.

To this objection, Colonel Bennett was quoted, by the Chicago *Daily News* of November 15, as saying: "There has never been an accidental nuclear explosion. The control devices are so good and so involved that an accidental explosion is not a danger." This sounds like a good, commonsense attitude, the voice of experience. Many military personnel get accustomed to living with dangers. A soldier knows that the grenade he carries could blow him to bits if the pin were accidentally pulled, but after living with it on his belt for a year he forgets about the slight danger. Even so, most civilians would prefer not to live on a powder keg without some very good reason for doing so.

Designers have worked hard to make the control devices as effective as humanly possible, and they must be good, for the record is very good. It even happens to be perfect. We don't hear much about the near-accidents, but in the case of one H-bomb dropped accidentally in North Carolina in 1961, it was reported that five of the six safety devices had failed. There were six, and the bomb was "unarmed" so there was no detonation. An H-bomb in the bay of an airplane can be carried "unarmed," with one vital part to be inserted before dropping, because there is plenty of time to "arm" it on the way to the target. Thus it may be intrinsically easier to make it safe than it is for a missile such as the Sprint, which must be ready to fire within a few minutes of the first warning and within a fraction of a second of identification of its target. We haven't had experience with those yet. But even ignoring this distinction, the good record is not completely convincing.

Experience with bomb accidents is the sort of stuff that the study of statistical probabilities is made of. Let us think about a variant of the ghoulish game of "Russian Roulette." A six-shooter has a cylinder with six bullet slots. Suppose you are given one not knowing whether it is loaded. You are permitted to spin the cylinder ten times—or even a hundred times—and pull the trigger. You do so and it does not fire. You are then to point it at your head and pull the trigger. Would you feel sure that you would not kill yourself? Fairly sure? Yes. But certain? No.

The armed forces have been storing or handling, let us say, ten thousand nuclear bombs for perhaps ten years. They point to the fact that none has exploded as proof that none will explode accidentally. They propose to store, at a guess, a thousand nuclear warheads near American cities for the next ten years. According to past experience, the probability that one of them will explode accidentally is not more than 10 percent. Citizens of Chicago may take comfort that that is divided among ten cities or so, so locally there may be only about one chance in a hundred of serious trouble in the next ten years. That is about all that can be proved by Colonel Bennett's reference to the good record. It may be good common sense to ignore a small risk like one chance in a hundred, even if the event would be catastrophic, for one feels that life is full of dangers. But let us look at the small chances on the other side of the coin.

Why are we installing this Sentinel system? The reasons are confused; they involve China and Russia, they involve military and industrial pressures on Congress, and citizen anxiety or apathy and many other factors. So let us simplify again by considering only the official reasons given for the Sentinel deployment. Colonel Bennett said: ''The Sentinel system is designed to defend the nation against a possible delivered missile attack by the Chinese Republic or an accidental launch of a nuclear-armed intercontinental missile by any foreign power.''

The same army spokesman who wants us to ignore the small chance of an accidental explosion at home by claiming that it does not exist is inviting us to worry about the chance that China, with a few missiles, will attack a country with thousands of missiles and to worry that an accidental launch of a Russian missile will hit one of our cities! There are few things of which one can be absolutely sure, but common sense should make us very nearly certain that the Chinese, at a time when they will have only a few intercontinental missiles, would not make a completely suicidal attack against the tremendous nuclear might of the United States. Such an attack seems much less likely than an accidental Sprint or Spartan detonation.

More serious than the Chinese ''threat'' is the technical possibility that an accidentally launched Russian missile might come our way. We have more than a thousand missiles in underground silos, with their computers and radars all adjusted to guide them toward various Russian cities and missile sites, and the Soviets likewise have several hundred missiles aimed at us. The chance that a Soviet missile would be launched accidentally may seem fairly remote. But what we are considering is more unlikely than that. We are considering the chance not only that a Soviet missile will malfunction and be launched, but that it will malfunction in such a way that it functions perfectly and aims directly at an American city eight thousand miles away. Although the likelihood of this double feat seems very small indeed, it is perhaps more probable than a Chinese attack.

Which, then, seems the more likely: one of a few hundred Soviet missiles being so perfectly launched accidentally as to hit an American city, or one of several hundred American missiles simply exploding accidentally where it sits on the edge of a city? The first seems to require two accidents in succession, the latter a single accident. Even if it is a fairly remote chance, it seems considerably more likely that an American city would suffer nuclear calamity from an accident at home than from a Soviet accident.

Thus, if the army persists in its plan to put the Sentinel missile sites on the edge of population centers, even from the limited local point of view the cure is worse than the disease. This situation could be remedied by moving the missile sites out into open country, where the Spartans would be just as ready to intercept an accidentally launched missile.

Civilians can make such a change when the army submits its missile-site plans for congressional approval, starting with a hearing before the normally cooperative Joint Armed Services Committee, scheduled for this month.

Introducing more danger than one is trying to prevent is typical of the whole effort to attain national safety through ABM defense. This larger folly can be remedied only by having the people and their Congress learn, perhaps through these local mistakes, that national safety is not to be sought by pursuing the will-o'-the-wisp of ABM defense. This defense would not be effective against a massive Soviet attack, according to those highly placed experts who have had a thorough look at the military and technical factors involved, but who have no vested interest in military empire building—former Defense Secretary Robert S. McNamara and all of the science advisers of the last three presidents. People must learn that national safety in the precarious nuclear age should be sought instead by more vigorous pursuit of international agreements—which the Soviet Union appears to be ready to pursue to our mutual benefit—by cutting off the deployment of offensive and defensive missiles of the nuclear giants, by avoiding the spread of nuclear weapons to many nations, and by otherwise "taming the atom" so that we may turn our energies more fully to improving the lot of mankind and removing the causes of war.

The Antiballistic Missile:
A Dangerous Folly

[1968]

In the next few weeks, the U.S. Senate is scheduled to take final action on a military appropriations bill that includes, among other things, funds for the deployment of the "Sentinel" antiballistic missile system. If the Senate approves the bill, the United States and the Soviet Union may well be on their way to a new round in the nuclear arms race. It would be a tragic decision not only because the Sentinel is dangerous, expensive, and unnecessary, but because it may jeopardize chances for further arms control agreements between the major powers.

Because appealing arguments can be made in favor of immediate deployment, the ABM has become an ideal vehicle for those who favor the perpetual expansion of our nuclear arsenal. ABMs are purely defensive in their immediate intent, and defense obviously seems virtuous. At its most superficial level, the argument is that, in an H-bomb attack of predetermined intensity, the Sentinel ABM system would save some lives. In the Senate debate of June 24, Senator John Pastore of Rhode Island, with polemic fervor, referred to the Soviet ABMs at Moscow intended to save Russian lives. Are American lives less valuable than Russian lives? he asked. If not, shouldn't steps be taken to protect Americans? Such oratory has a plausible ring. A conservative senator, admittedly confused, can tell his constituents that, if he should be wrong, he would prefer to err on the side of safety. But since much more goes into determining which is the safer course, it seems far more likely that lives will be saved by not acquiring ABMs.

The intensity of a potential attack is not fixed. Both the likelihood of an attack and its intensity would probably be increased by our decision to deploy ABMs. The natural Soviet response to an American ABM system would be to build more missiles to penetrate it, just as we have already responded to their small ABM deployment around Moscow by adding more deliverable warheads. The Russians would also probably deploy a serious ABM system in response to ours; we would respond to that; and so on in the upward spiral. Intensification of the arms race increases the chances of nuclear war.

It can be argued that by deploying ABMs we are less apt to be attacked since we have a shield to defend ourselves, and that the development of a perfect ABM defense would effectively eliminate the threat of nuclear war. Such arguments assume that a good defense is possible against nuclear weapons. It is not, and it probably never will be. But since one cannot be sure, those who believe we should not deploy ABMs at this time still favor continued research to determine whether an unexpected breakthrough could make possible an effective shield in the future. Premature deployment merely detracts from this effort by misdirecting talent and funds.

The proposed ABM defense system involves exploding our own H-bombs over our own cities, in the hope of intercepting enemy ICBMs there. The Sentinel system has two parts: one based on long-range Spartan missiles that explode several hundred miles out, the other the short-range Sprint missiles that are designed to explode more than ten miles and preferably as much as fifty miles above our cities. There is some uncertainty what damage they may do. They will presumably be at least ten times as powerful as the early A-bomb that destroyed Hiroshima from half a mile up. In response, the potential attacker will try to make his more powerful attack missile explode where it is intercepted, even if it fails to get through. This is, of course, very much better for us than having it explode lower down, but it is hardly a perfect defense.

The much more serious technical objection to any ABM deployment is that it cannot stop more than a modest fraction of the incoming missiles in a massive attack. Because of this, an attacker can undo the effectiveness of the ABMs by attacking with more missiles. When he was secretary of defense, Robert McNamara testified that the response of mounting more attack missiles would cost less than the ABM system. He also quoted all of the science advisers, past and present, to Presidents Eisenhower, Kennedy, and Johnson, as well as former directors of research of the Department of Defense, as being opposed to mounting ABMs against a possible Soviet attack. Almost all of these men have since declared themselves opposed to any ABM deployment at this time. These are extremely competent men, not professionally biased, who are in the best position to appreciate the technical inadequacies of ABMs.

The ABM system can never be tested in advance under actual conditions and it will remain quite uncertain how much it can do. It is considered conservative military practice to prepare to cope with overestimated enemy capabilities. With all the uncertainty of ABM effectiveness, the likelihood is that attack intensities would be increased more than enough to compensate for ABMs on both sides. As a result, the anticipated devastation would be greater than before. The senator who thinks he may be erring conservatively on the side of safety by voting for ABMs must hope either that the Soviets won't notice or that we can overwhelm their production capabilities, as we used to believe we could in the pre-Sputnik days. As Mr. McNamara and other officials have emphasized, both sides will have misspent a lot of effort and money with no increase in security. The added factor of uncertainty of performance will probably even decrease security by making the deterrent posture—the "balance of terror"—less stable and thus making nuclear war between the nuclear giants more likely on some hypothetical accounting sheet.

All this accounting must be kept in perspective. We are now in the realm of asking which side can destroy the cities of the other more times over. If, for example, we can destroy their cities five times over and they can destroy ours perhaps twice, these gains or losses in relative deterrent strength don't matter very much. We will presumably remain well ahead of the USSR in numerical nuclear strength for a long time, and they may be willing to let us. But even this is not extremely important. Mr. McNamara has assured us that we must consider ourselves adequately deterred from striking the USSR by their deterrent strength when we have four and a half times as many deliverable warheads as they have. They would presumably be similarly deterred even if the ratio were the other way around. Approximate equality is quite adequate. It may be impossible to stop the arms race if one side insists on staying too far ahead.

The continued arms race, with competition between defensive and offensive efforts, increases the likelihood of nuclear war in a more important way through the tensions engendered by the race itself and by the need to foster continued citizens' enthusiasm to foot the bill. If both societies would instead make efforts to more adequately meet their civilian needs and the needs of poorer nations, while desisting from ABMs and letting the arms race taper off, peaceful coexistence would be more firmly established.

Thus even if we completely discount the possibility of obtaining benefits from formal international agreements, starting an ABM competition with the Soviet Union by starting the Sentinel deployment at this time seems to be very bad policy. And in view of the possibility of useful arms control agreements—with which it would interfere—it is utter folly. Two types of international agreements now pending would be adversely affected by our decision to go ahead with ABM deployment this year—the Nonproliferation Treaty and an agreement with the Soviet Union on the limitation of offensive and defensive nuclear weapons. Both have been patiently promoted by our government over the last two years and both are now at a hopeful stage of progress that should not be interrupted because the left hand in the Department of Defense ignores the right hand in the State Department. Further agreements would be indirectly affected, too, and the 1963 partial test ban treaty, the successful prototype of mutually beneficial nuclear restraint, would be in danger of collapse. There would be new demands for the resumption of detonations in the atmosphere to test ABM systems and a new danger of fallout.

The Nonproliferation Treaty, on which the United States and the USSR finally concurred last spring, has been approved by the UN General Assembly and needs to be signed and ratified by as many nations as possible if it is to be effective in helping to prevent the spread of nuclear weapons.

The treaty, while in the interests of all, is necessarily somewhat unfair because it demands that the nations not possessing nuclear weapons give up their sovereign right to acquire them, while it makes no such explicit demand for restraint by the nuclear powers. Furthermore, it has an added value to the nuclear powers in that it preserves their special position. The world must live with the existence of a few nuclear "haves" and many "have-nots," and the best the treaty can do as a first step is to preserve this situation. The nonnuclear nations naturally tend to resent the one-sided demands. As an inducement for them to sign and to continue to abide by the treaty, the nuclear giants agreed to include in the treaty, as its Article VI, a declaration that they will negotiate steps of arms limitation in good faith. No specifics were incorporated in the treaty itself because, it was argued, it would take too long to agree on the details. We cannot expect to delay proliferation very long unless this implied promise is taken seriously. Both of the great nuclear powers have good reason to avoid proliferation and thus should feel strong pressure to negotiate further steps of arms restraint. ABM deployment moves in the opposite direction and will be hard to stop if we let it start.

Negotiations to limit offensive and defensive nuclear missiles are still in a preliminary stage. The USSR, probably because of gruesome memories of the invasions of Napoleon and Hitler, has a special interest in defensive weapons. The Soviets undertook the premature installation of ABMs around Moscow not long after President Kennedy decided against making the mistake of deploying the old Nike Zeus system. Mindful of our long lead in strategic attack weapons, the Soviets refused to restrict weapons discussions to defensive missiles. For the past two years, President Johnson has been proposing discussions aimed at limiting both offensive and defensive weapons, and finally, last June 26, the Soviets responded favorably to this proposal. The slow wheels of diplomacy are now starting to grind out arrangements for negotiations.

With the pressures arising from Article VI, as well as the obvious advantages of nuclear restraint between the giants, there is every reason to believe that the

Soviets are sincere in their desire to agree on limitations. But the negotiators are undoubtedly in for some hard bargaining over details. Each side will want to maintain pressure on the other in the form of readiness to resume the arms race if the negotiations fail. But it is just as important that each side make it clear that it will retain, in its internal political process, the option to discontinue arms procurement and, in particular, the deployment of ABMs if the negotiations should succeed.

McNamara warned in a speech near the end of his career as secretary of defense that there is a "mad momentum" inherent in the procurement of new nuclear weaponry. How can the negotiators, particularly the Soviet negotiators, find it plausible that we could ever stop short of completing the "Sentinel" system, at least, if we cannot stop tentatively at a time that is as propitious as the present?

Our ABM decision has been confused further by putting our carefully nurtured Chinese ogre in the picture. Since experts agree that ABMs against the Soviets make no sense, the administration last September gave in to congressional and other political pressure and announced its decision to proceed with the Sentinel system, for which funding is still in question, as a "Chinese-oriented" system. While any system we could build would be like a sieve to a massive attack, this so-called light system would presumably shield cities from a few very simple intercontinental missiles. This might be the maximum capability of the Chinese until the mid-seventies, if they should be so implausibly obliging as to overlook using fairly simple penetration aids, such as "chaff," to confuse radar. But there are other ways the Chinese could evade the system—by launching missiles from near our coasts, for instance. Yet it is preposterous to worry that the Chinese might be so mad as to make an utterly suicidal attack against our vastly superior nuclear might. An ABM system, even if it were effective for a short period, should not appreciably increase our confidence in our ability to throw our weight around in Asia—which is probably what "Chinese-oriented" really implies.

Making nuclear war less likely by stopping the arms race is the most important reason for not deploying ABMs, but there are many other considerations that complicate and, on the whole, strengthen the decision. (1) It is argued that if a Soviet missile should accidentally be launched toward one of our cities, we could intercept it with ABMs. Though possible, this is improbable; and with continued proliferation and arms racing there would be more missiles that could be fired accidentally. It is also possible that our own ABM warheads might accidentally explode in our own cities. (2) Deployment of ABMs by the nuclear giants, it is also argued, would make it more difficult for newcomers to mount a significant nuclear threat—it would raise the entry fee to the nuclear club. But for them, challenging the giants is hopeless anyway, and their main incentive is to be able to challenge or to deter each other. (3) The proposed expenditures on ABMs represent fiscal irresponsibility in a year when budgets are strained, foreign financial centers concerned about the dollar are watching us for signs of restraint, and the crying needs of our convulsive cities and of the world's poor are being neglected to swell military budgets. (4) A year's interruption in the Sentinel schedule, it is said, would increase the cost of completing the system, but the negotiations thereby fostered might succeed in making completion unnecessary. If not, the delay might permit improved design and hence less rapid obsolescence of the system.

In the successive Senate debates on April 14, June 24, and August 1, at various stages of the military construction and appropriation bills, the Sentinel system seems to have become more of a symbol than a system of defense, a symbol of a determination that we must stay strong by continuing the arms race, come what may. Support came indirectly from elements in the "military-industrial complex" and directly from the Department of Defense. A substantial minority of senators, led by Senators Cooper, Symington, McGovern, Hart, Nelson, Clark, McCarthy, and others, supported amendments aimed at delaying initial ABM deployment for a year by deleting a few hundred million dollars here and there from this year's appropriations. However, the majority in the Senate was opposed to delay. The senators who supported the ABM argued from shifting grounds. The ABM system was at first purely Chinese-oriented—and the Soviets could be expected to understand this and not respond—until it appeared that Chinese missile progress had been delayed so we could afford to delay our Chinese-oriented system. Then, in the June 24 debate, it was advocated as a first step toward a massive Soviet-oriented system, partly on the grounds that the Russians had ignored our proposal of talks to limit offensive and defensive missiles.

Two days after that, the Soviets announced that they would talk. The proponents then argued that we should continue deployment on schedule to put pressure on the Soviet negotiators—and besides, they said, it was Chinese-oriented.

In June, the Senate vote most nearly favorable for delaying deployment was 51–34 and in August it was 46–27, with a vote of similar proportions in the House. It is unusual to have more than a token two or three votes in opposition to any part of a military appropriations bill, and the vote in June seemed encouraging, considering that the Soviets still would not talk. More senators could have been expected to vote for delay when the Soviets said they would talk. It is discouraging that the lighter August 1 vote, taken after favorable debate (but rather hastily when some members were off to the Republican convention), did not show such a trend.

The final opportunity for Congress to stop this year's ABM appropriation occurs when the military appropriations bill comes up for final vote, probably in the next few weeks.

For many reasons—for the sake of avoiding further dangerous upward spirals and outward spreading of the nuclear arms race, for the sake of a trend toward sanity in international relations, for the sake of fiscal responsibility and adequate attention to the nation's and the world's poor, for the sake of an all-around safer world in the future—it is greatly to be hoped that more senators will react to the Soviet decision to talk and will appreciate that the best way to try for successful negotiations is to keep our options open and to avoid deploying ABMs prematurely and uselessly now out of sheer momentum.

5 Disarmament Procedures

Shortly after the first chain reaction in a reactor (then called a "pile") at the University of Chicago in late 1941, the Argonne Laboratory was established in Argonne Forest, with a ten-year lease from the Chicago Park District, under duress of wartime, to continue the work with "Chicago Pile II." My discussions with Donald Flanders, leading to the first of the articles in chapter 1, "A Deal before Midnight?" were held in the sylvan atmosphere of the old lab in 1949 and 1950, before the move to a new site when the lease was up. Later, in the commodious new laboratory, Drs. Arthur Jaffee and Melvin Freedman joined us, and our little discussion group was thus expanded to four regular members with others participating occasionally. U.S. disarmament negotiating initiatives, beyond the Acheson-Lilienthal-Baruch plan, seemed to us to have been hastily prepared and slanted in our favor but had convinced most of the public that reasonable approaches had been tried and that the Soviets would not agree to anything sensible. It was to counter this impression that we felt impelled, after considering and rejecting several "harebrained ideas," as we called them, to write in some detail our "specific proposal" that we felt had a better chance of acceptance and of contributing to national security than any of the initiatives that had been tried. The military situation was unbalanced and the negotiation proposals showed a corresponding bias. The Soviets wanted to reduce the numbers of atomic weapons in which we were ahead and we sought to limit conventional forces with which the Soviets were more massively equipped.

In our specific proposal we emphasized the possibility of balancing the concessions in the two categories. Being physical scientists, we dealt in numbers. We felt it useful to quantify concessions for the sake of bookkeeping, even though the evaluations could be made only roughly. The detailed listings that appear there seemed more appropriate in those earlier times when stockpiles were new and small than they would now, but the general principle of balancing concessions in different categories of weapons is still relevant in our era of multiple overkill.

We chose not to publish our proposal but instead circulated it internally, hoping it might have a greater chance of influencing negotiations if un-published. Ten years later, when it had no more than historical value, it was published with an accompanying article comparing it with other disarmament proposals that had arisen in the private sector. The specific proposal and the accompanying article, dated 1952 and 1962, appear as the first and third articles of this chapter.

During the late fifties the prospects for arms limitation were dimmed by the advent of H-bombs delivered by intercontinental missiles, and the secrecy of launching sites became of prime importance to the Soviets. Secrecy versus inspection became the problem. It was said that the Soviet negotiators wanted disarmament without inspection and the Americans wanted inspection without disarmament. The region-by-region idea, with disarmament and inspection introduced systematically in one region after another, cropped up indepen-dently from both the Soviet side and the West at the Pugwash Conference of Scientists on World Affairs held at Moscow in 1960, the first of several Pugwash conferences that I attended. This approach is described in the second of these articles, dated 1964, and elsewhere. It seemed to me to be an impor-tant possibility, if the leaders of the two sides had wanted a way to reverse the arms race safely despite mutual distrust. That article also discusses how one limited agreement, such as a test ban, might build confidence to facilitate a next agreement, and so on, a less demanding procedure than the other schemes that require agreement in advance on a sequence of steps. In a sense, this is what has happened, the agreements being the partial test ban, the nonproliferation treaty, and an agreement against ABMs, but it has happened much more slowly than was envisioned and is of more limited scope. While mutual arms reduction has been the subject of much official rhetoric, it has never been given the priority over arms procurement that would be needed for the adoption of a systematic disarmament plan such as here discussed. In-stead, each small step has involved a struggle between advocates of limitation and those pushing procurement.

A Specific Proposal for Balanced Disarmament and Atomic Control

[1952]

The disarmament plan presented here is intended to provide a demonstration that United States diplomacy has not yet tried all the approaches that should be tried to the problem of negotiating an agreement to end the atomic armament race before it gets completely out of hand. In this sense it is a sample plan. As a sample plan, we believe that it is more realistically representative of the needs of our nation and of the world today than are such predecessors as the report of the Acheson-Lilienthal committee, which was applicable to the needs of 1946, or of the recent sweeping "Department of Terrestrial Magnetism" proposal, which departs, with somewhat more reckless abandon than seems politically feasible, from our professed desire to base atomic disarmament on the best available guarantees against evasion.

But in another sense it is more than a sample plan. It is a serious proposal for an outline of procedure which, while perhaps not unique, cannot, we believe, be paralleled by many plans of equal merit. This plan is perhaps not only getting close to the spirit of negotiation most compatible with our long-term national needs, but also to the detailed plan which should be made an important part of our foreign policy if we are to maximize the probability of avoiding a future atomic impasse. The disarmament problem is so thorny and complex that we can here do little more than present an outline and a general justification for this as a basis of procedure: the details are sketchy and tentative, being in need of more accurate assessment by specialists in various fields. Because this plan may be getting close to a usable form, we are giving it a closely restricted initial circulation, for international feelers may perhaps be more effective when not spoiled by propaganda effects.

In spite of the frustrations of previous negotiation, of which we are fully aware, in spite of the evasiveness of the Soviet substitute for diplomacy, we conclude that we should continue to advocate further efforts toward effective negotiation. We conclude further:

(1) that it is to the mutual advantage of the U.S., as representative of the free world, and of the USSR, as representative of the Soviet orbit, to minimize the risk of future destruction that amounts almost to mutual

---

Coauthored with D. A. Flanders, M. S. Freedman, and A. H. Jaffey.

annihilation by agreeing on a disarmament plan which balances the advantages and the necessary concessions for the two sides;

(2) that such a plan has not yet been proposed to ascertain whether the USSR would agree to it;

(3) that even after a control plan provides complete access to and inspection of atomic production facilities, there will remain an uncertainty in the amount of crucial (fissionable and fusionable) material that each side has reason to believe the other has produced, expressible roughly as a percentage (which we take to be somewhat less than 20 percent of the total past production);

(4) that because of this uncertainty and the consequent possibility of hiding a secret stockpile withheld from the disclosure of past production, it is reasonable to expect cautious and distrustful parties to an agreement to be willing to reduce their stockpiles in a controlled manner only down to an agreed-upon ceiling equal to some fixed percentage of the *larger* atomic stockpile (which we take to be about 20 percent, in keeping with the uncertainty in verification of past production);

(5) that the rapid growth of stockpiles and the consequent growth of the actual amount of uncertainty in past production places a great urgency on stopping production of atomic materials soon, particularly on the side having the larger atomic stockpile;

(6) that in order to be acceptable and to operate in the present distrustful world atmosphere, a realistic disarmament scheme must consist of stages, the accomplishment of no one of which would give either side an appreciable advantage in the hypothetical event of a breakdown in carrying out the agreement or of a subsequent outbreak of hostilities; and

(7) that there exist two sorts of asymmetry in the present alignment of world power (a preponderance of atomic armament on the U.S. side and of armed manpower on the USSR side, and the greater degree of information available to the USSR about the U.S.), which would make it disadvantageous to one side or the other if precisely the same steps of disclosure and disarmament were to be taken simultaneously by both.

Because of the uncertainty whether an atomic stockpile would really have been eliminated when it was supposed to have been eliminated, we propose a plan whose explicit goal is partial rather than complete disarmament. Both atomic and conventional arms and armed forces are to be reduced to agreed-upon ceilings, compared with which the uncertainty will not loom so large as to seem potentially decisive. The plan includes a complete schedule of full disclosure and of disarmament down to the agreed-upon ceilings. In the specific example we present, the atomic ceiling is taken to be 20 percent of the larger stockpile, while the ceiling of "conventional" arms and of armed forces is taken to be roughly one-third of the initial levels. The rather high ceiling for conventional arms, where more nearly complete disarmament might seem to be a more desirable goal, is chosen so that a still rather substantial conventional armament may be relied upon to help reduce the importance of the atomic uncertainty. Thus the atomic stockpiles need not be kept large enough to do this alone. It is envisaged that the entire plan will be agreed to in advance of the effecting of any of the stages.

Once these immediate and scheduled goals have been attained, there would be expected to follow a probationary period during which the ceilings would remain fixed. In this condition it does make sense to await an improved international climate before hoping to undertake further steps of disarmament, whereas with an atomic arms race in progress, it does not. The half-life of tritium is about twelve years, so the uncertainty in this part of the crucial-material stockpile will shrink. Guarantees against clandestine production must be good enough to assure that uncertainties in the rest of the stockpile do not appreciably grow. The long-term cooling of tempers may make the uncertainty shrink in apparent importance. Then further disarmament may be expected.

What is treated in this article is the approach, by stages, of atomic and conventional disarmament, without corresponding elaboration of the necessary continuous control and/or inspection (defined here to include complete and unhampered access except as limited by prior agreement) following completion of disarmament. We believe that no extension of methods and procedures beyond those necessary to achieve the stages in this plan would be required for a high degree of assurance in such an inspection system. Since the final result of disarmament involves the retention of a substantial armament, the stringency of inspection need only assure that the maximum clandestine production is small in comparison.

In order to go about systematically designing a stage-by-stage plan in which neither side gains an appreciable military advantage over the other through the carrying out of any particular stage, it is desirable to assign, by the best method of estimating available, numerical values which will be taken to represent the loss in relative military potential represented by each step of disclosure or disarmament conceded by each side. Once the steps are listed and evaluated, they are arranged in the order in which they should be carried out. For simplicity of negotiation we make this order the same for both sides, though the stages at which any particular step will be carried out will be different for the two sides. The order is selected so as to accomplish the most desirable first things first. In particular, cessation of production of atomic crucial materials in declared plants is put as early as it can be and still be verified by minimum access.

We arbitrarily select seven as the number of stages. Having evaluated and ordered the steps, we next divide them into stages in such a way that the value of the total concession by each will progress toward the goal in seven jumps of approximately the same size, but with the atomic part of the jump larger than the conventional-armament part for the U.S. in accordance with the greater evaluation of its atomic facilities, and the other way around for the USSR.

In this discussion we use the abbreviation "U.S." to mean "United States and the rest of the free world," or "West," and "USSR" to mean "Russia and its satellites," or "East." It is considered very likely that, if agreement could be reached by the two principal contenders in the atomic armament race on the general nature of a plan such as this, the other nations associated with them would be only too glad to go along, so the abbreviations used are suggestive of the principal parties to the negotiations. The evaluations include the facilities of the associated nations.

In assessing a value of each category to each side, "value" is rather loosely defined but is related to the following assumed immediate aims (which might be considerably modified by successful disarmament). It is assumed that the military aims of the United States are to be in a position to help other nations of the free world effectively to resist Soviet pressure and prevent Soviet invasion, to minimize the effectiveness of a possible Soviet atomic attack both at home and abroad, and to maximize the prospective effectiveness of a threatened atomic attack on the Soviet orbit. All of these aims center around the overall intention of minimizing the temptation of the Soviet leaders to initiate further expansion by making it appear unprofitable. It is assumed that the military aim of the Soviet leaders is essentially the opposite counterpart of this, the overall intention for the immediate future being not only to maintain a tenable defense, but also to stabilize and, if possible, to extend the Soviet orbit, and to influence the rest of the free world into doing nothing about it. Insofar as the military aims of the two sides are the opposite counterparts of one another, the "value" of a given disarmament step by a given side should be roughly the same from the point of view of either side viewed as a gain by one and a loss by the other, so we do not specify which point of view is taken in defining "value." Our estimates are rough, so the distinction is not very significant. If there is doubt, we try to strike an average of the two points of view, so as to arrive at a plan which will come as near as possible to looking equitable to both sides, so far as we can judge.

Even though the numbers assigned represent nothing more than subjective judgments concerning relative values, and so can have no very exact meaning, they are useful in helping us in a systematic way to arrange balanced stages. The numbers are intended for the use of American advisers and policymakers only and they are not intended to appear in any proposal to a foreign nation unless it is deemed that this would help ''sell'' the plan in the negotiations. It seems to us more likely that the numbers would supply needless grounds for international bickering, so the result we arrive at by the use of the numbers should be translated back into words in writing a diplomatic proposal. The numbers will then have served to help convince us that the plan is equitable. We shall here describe our stages both in terms of the numbers and in words.

The steps to be evaluated are divided into the main classes ''atomic'' and ''conventional.'' The atomic class includes geographical access to territory because this access in detail is more important to the success of atomic verifications of declarations than it is to the ''conventional,'' though it is of course important to both. The ''conventional'' class includes all nonatomic categories, even though biological warfare and other advanced weapons are not properly described by the epithet.

For the sake of brevity in the evaluation tables (tables 1 and 2), we use a code to name the various steps of disclosure and disarmament in the various categories of armaments and facilities. The various *atomic* steps are denoted by *lowercase* letters, with subscript numbers to indicate the categories. The *conventional* steps are denoted by *capital* letters, with similar numerical subscripts to indicate the categories.

Our reasons for evaluating the steps as we have are incomplete because of limitations of time and access to information. A more careful reevaluation would be desired if it were to be used as a basis of actual negotiation. Even our incomplete reasons cannot be described completely here, for our discussions have been longer than this report. We shall, however, now attempt to list the most relevant considerations which have influenced the judgments represented by the numbers assigned to the various steps. Values we have assigned are given at the end of each section.

Reasons for Evaluation of Steps, Atomic:

a_{234} Declare stocks and present production rates of uranium, fissionable material, and atomic weapons. U.S. declaration has moderately small value, since strategic value of knowledge lies in the fact that we have a

TABLE 1. Evaluation of *Atomic* Steps

Step	Value of concession by	
	U.S.	USSR
a_{234}	4	9
be_{12}	2	4
beh_2	18	18
fgh_2	3	4
b_4c	2	12
e_4d	3	15
f_3	6	6
h_4	6	6
$afgh_1$	4	5
g_3	18	17
fg_4	35	16
bfg_5	10	6
i	154	56
Total value	265	174

Asymmetric plan—code and evaluation for *atomic* arms and facilities
a. Declare inventory of the products of:
b. Declare location of:
c. Permit search of territory by aerial survey, with declaration of use of installations spotted, as requested.
d. Permit ground access to entire national territory (perhaps with specified, small-area local exceptions), with similar declaration of use.
e. Permit exterior access to:
f. Permit interior access to (in sufficient detail to show exterior of machinery):
g. Permit complete access to, including product sampling, checking of records (and personnel questioning), and dismantling of apparatus where necessary for verifying declarations. Samples to be dismantled are to be selected by inspection team but limited in number to remain within scheduled cutbacks of production or stockpiles:
h. Discontinue production of:
i. Eliminate (weapons, materials, and production facilities) down to a previously agreed-upon ceiling. This item is listed fractionally: for example, i(20%) means eliminate to the extent of 20% of the way from the starting point down to the ceiling. *i* means the same as i(100%):

The categories to which these steps (except c and d) apply are indicated by subscripts:
1. Mines
2. Refineries
3. Production plants for materials (fissionable and fusionable)
4. Weapons production installations
5. Crucial materials and weapons

TABLE 2. Evaluation of *Conventional* Steps

Step	Value of concession by U.S.	USSR	Combined Step	Value of concession by U.S.	USSR
A_1	3	10	A_{13}	6	20
A_2	4	10	A_{56}	20	25
A_3	3	20	BC_1	5	17
A_4	6	0	A_7	20	20
A_5	10	5	B_3	3	10
A_6	10	20	A_{24}	10	10
A_7	20	20	B_{56}	11	10
B_1	1	5	B_{24}	9	5
B_2	2	5	C_{24}	16	20
B_3	3	10	C_{56}	24	30
B_4	7	0	C_2	7	15
B_5	7	5	BC_7	23	23
B_6	4	5	D_1	6	20
B_7	20	20	D_{34}	24	30
C_1	4	12	D_{256}	46	66
C_2	7	15	D_7	5	5
C_3	4	20			
C_4	12	0	Total value	235	326
C_5	10	5			
C_6	14	25			
C_7	3	3			
D_1	6	20			
D_2	10	22			
D_3	6	30			
D_4	18	0			
D_5	15	7			
D_6	21	37			
D_7	5	5			
Total value	235	326			

Asymmetric plan—code and evaluation for *conventional* arms and facilities

A. Declare location of fabrication and development installations, permit exterior access, and discontinue production of:

B. Declare inventory, supply construction (organization) plans, and permit detailed inspection of:

C. Eliminate half-way down to prescribed ceiling of:

D. Eliminate rest of way down to prescribed ceiling of:

1. Armed forces personnel (uniformly by categories)
2. Tanks and artillery
3. Submarines
4. Naval surface craft
5. Planes, long-range
6. Planes, short-range
7. Rockets, guided missiles, B.W., and other advanced weapons.

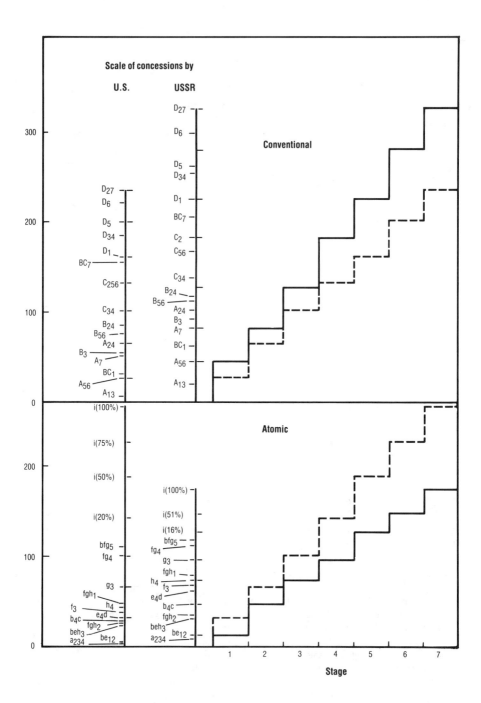

Fig. 1   Assymetric plan — separate schedules of concessions

preponderance, not in the exact size of U.S. stockpile, and the Russians probably know it pretty well anyway. USSR declaration has more value because its stockpile is probably small enough that its size is crucial for its effectiveness, and we know less about it. (U.S.–4; USSR–9). . . .

[The preceding paragraph is a sample to give the general idea of the evaluations. Its result appears as the first entry at the lower left of figure 1. The meaning of the symbol a_{234} and similar symbols in the figure should be clear from the codes listed below Tables 1 and 2. Similar explanations for the many other steps and much further detail and discussion in the original paper are here omitted.

Several such steps are included in each of the seven stages, as can be seen in figure 1, where each stage appears as one step of a stairway for each nation (U.S., broken lines; USSR, solid lines). Each stage is to be carried out in a specified time of perhaps a year or so. The total evaluation of all the steps in each stage is indicated in table 3. It illustrates the idea of keeping the total concessions made by the two sides equal at each stage, balancing atomic and conventional concessions.]

Stage I

U.S. Atomic. Declare stocks and present production rates for uranium, fissionable materials, and atomic weapons; permit exterior access to mines and refineries, followed by detailed disclosure and shutting down of refineries; discontinue production of crucial-material producers, ver-

TABLE 3. Evaluation of Stages

Stage	Atomic Concession by		Conventional Concession by		Total Concession by	
	U.S.	USSR	U.S.	USSR	U.S.	USSR
1	32	13	27	45	59	58
2	34	34	38	37	72	71
3	35	27	36	45	71	72
4	41	22	31	55	72	77
5	46	31	29	43	75	74
6	38	20	39	55	77	75
7	39	27	35	46	74	73
Total value	265	174	235	326	500	500

ified by exterior access; declare location of weapons-producing and development installations; and permit aerial survey of all territory.
U.S. Conventional. Declare location of soldier training centers, permit
exterior access, and discontinue soldier training; declare location of development and fabrication installations for submarines and all planes;
permit exterior access to these and cease all production.
USSR Atomic. Declare stocks and present production rates for uranium,
fissionable materials, and atomic weapons; permit exterior access to
mines and refineries.
USSR Conventional. Same as US.

The order of the steps as well as the magnitude of the stages is influenced
primarily by the requirement that the plan shall be operable in spite of the
present high degree of international distrust. The initial stages will presumably be performed hesitantly by both sides, each watching the other to learn
whether it intends to carry out the obligations of the plan. Of the items of
cessation of production, disclosure, and disarmament, some are less completely irreversible than others, and those which involve the least irretrievable
loss of military advantage seem most appropriate for the early stages. The best
example of such a step is cessation of production, for, in case of a later
breakdown of the plan, production can be revived again with loss only of the
material that would have been produced in the meantime. Compared to this,
elimination of a large stockpile down to parity with a lagging competitor has a
much greater finality and is therefore put off until the last few stages. Because
the effectiveness of the plan to achieve disarmament is limited by the size of
the larger atomic stockpile, cessation of the production adding to this stockpile is halted in the very first stage, with a minimum of concessions of
disclosure. Exterior access to declared production facilities is considered sufficient verification at this stage. The access for search of territory which
makes it possible to verify that all production facilities have been declared
involves much greater permanent loss of information and is put a little later.
Detailed inspection of the facilities makes sense only after this assurance is
available that all facilities are known. . . .

Because the uncertainty in the verification of past production limits our
goal to partial disarmament, the strict control of current production provided
by the international-ownership mechanism of the Baruch plan appears disproportionately stringent.

A related question is whether power piles would or would not be permitted, a question on which the Russians have made their views clear in the past.

Obviously no workable agreement would allow for the unlimited construction of power piles, especially in view of the possibilities for breeding. After the completion of the stages in the above plan, further production of fissionable materials might be limited to the small amounts produced in low-power research reactors. If power piles are to be permitted by the agreement, their source materials might consist solely of the stockpile remaining legally within the possession of each nation. In effect, this is to say that there can be no substantial development of an atomic power industry. Should unlimited development and construction of power piles be insisted upon, we do not see how the production of fissionable material can be kept small in comparison with the stockpiles allowed. . . .

Thus, while the retention or production of even small amounts of fissionable material by individual nations was formerly considered intolerable, it is no longer possible to restrict this amount to less than the order of perhaps one to two hundred bombs equivalent (using published estimates as a guide to U.S. production). Enigmatically, it is our production which continues to elevate this figure. Thus it is for the best interest of the U.S. to pursue an agreement to stop further production before the irreducible limit of error exceeds reasonable bounds, and at as early a point in the necessarily lengthy time schedule of stages as possible.

In offering the Baruch plan to the UN, the United States made a gesture unparalleled in history by proposing to give up a dominant military force when it had sole possession and the prospect that its advantage would not be immediately dissipated. While this kind of opportunity has irrevocably passed, we are still able to give strong evidence of our sincerity and desire for a settlement by offering to reduce our undoubted advantage in the field of atomic weapons at a greater rate than is required of the USSR, compensated by a corresponding greater rate of disarmament in conventional weapons in which, at least in regard to manpower, the Russians have superiority. Further, the fact that Russia has the atomic bomb, and almost certainly a more detailed knowledge of our facilities, past production, and present production rate than we have of theirs enables the U.S. to offer verification of this information at an early stage without substantial loss in military security. Thus, we may "dramatize" our sincerity with concrete action, secure a worthwhile concession from the Russians on conventional arms, and still at each stage have a test (in terms of Russian compliance with the terms of the agreement) of USSR willingness to continue the stages of disarmament, without a disproportionate loss in relative military potential.

A prime consideration in our deliberations is related to our concern that an agreement must be acceptable to the Congress before being put into effect. Although it is outside the province of this report to deal with the political methods of negotiation or presentation of a plan, we believe that a significant gain would be realized if some assurance of the support of Congress for the general principles and spirit of the plan could be obtained prior to its presentation to the USSR. With such backing our representatives in the discussions would have clearly before them the limits within which settlements of differences could be compromised and would not be operating in a vacuum, uncertain as to U.S. acceptance of the agreement. The difficulty here arises that Congress does not keep secrets so the negotiation almost inevitably becomes prematurely confused with the "propaganda war." Congressional concern about the magnitude of the atomic concessions offered in this plan by the U.S. may be tempered by the arguments that most of the information-granting steps are, in large part, already in fact in the hands of the USSR, and that, as we have a stockpile which is approaching saturation, the loss in current production occasioned by cessation of production represents a marginal change in U.S. military potential, while the corresponding concessions by the Russians substantially reduce their atomic striking power.

The proposal must be carefully designed to avoid presenting damaging propaganda setups to the Russians. It is probably not possible completely to eliminate this factor. Negotiations carried out in private minimize the opportunities for propagandizing and thus may be recommended. A plan which offers both the main feature of the Russian proposal—immediate cessation of production—and the asymmetric approach to equality in which our steps toward atomic disarmament are larger than theirs furnishes a more reliable and obvious test of ultimate Russian intentions by putting on them the onus of rejection.

Step-By-Step Disarmament

[1964]

Disarmament is an avowed goal of the foreign policy of both the United States and of the Soviet Union and has been during the buildup of more than half of the armed might of both countries. There are several reasons why the avowed goal is not more vigorously pursued. One widespread opinion is that we cannot have disarmament because the Russians will not agree to a sensible plan. Until recently, a similar opinion was held of the nuclear test ban, yet we did come to a useful agreement on that. This opinion suggests that the "avowed goal" might mean no more than that each side wants disarmament if it can get it on its own terms. During the long and dreary sequence of disarmament negotiations, the terms of each side have typically been such as to perpetuate a superior position in the arms balance or to strengthen a weaker position. That is, the terms have had in most cases a short-term motivation—a one-sided military advantage for the near future. Yet there have been sincere attempts to narrow the gap.

Most political policies and most political decisions have a short-term motivation, and are typically aimed at solving the immediate and pressing problems in time for the next election. If we look at only the next few years, we have only one other powerful nuclear nation to worry about. It appears militarily practicable for us to maintain superior strength, so this may seem to be the most straightforward and safest way to discourage attack against us. Furthermore, seeking safety through military strength is the traditional course, inherited from the prenuclear age when there was much less reason to avoid war.

Political decisions that run counter to established custom are very difficult and are made only as the result of a great crusade or inspired leadership. It is much easier to pass a budget appropriation for an established department than to establish a new one. It is also difficult to defeat an appropriation for established activities upon which many political leaders and their constituents are dependent for support. Thus, the fact that we continue on a course of ever-increasing armaments is not necessarily proof that this is the wisest or the only practicable course.

The nuclear age has brought a new need for disarmament. The fact that disarmament negotiations failed to achieve much before World War II has little bearing on present prospects—the need and the incentives for achieving a new system of world stability are now so much greater than they were. The fact that similar negotiations have failed since World War II seems to mean that the nations have been slow to react to the new realities and the growing urgency.

Faced with the need to make nuclear war less likely, some people favor arms control as a way to perpetuate national military strength, some favor arms control as a step toward eventual disarmament, and some favor vigorous efforts to achieve disarmament, with adequate provisions for world stability, as soon as possible, perhaps with some incidental arms control if it can be negotiated in the meantime.

Small Step Chain Reaction

There is surely something to be said for trying to negotiate one small step of disarmament after another. The already successful negotiation of the partial test ban treaty has improved the international atmosphere enough so that it should be easier than it seemed earlier to negotiate a further treaty banning underground nuclear tests. An agreement to ban underground tests might make it easier to negotiate a treaty banning the production of missiles, with adequate inspection, and so on. The hope for a sequence of such small successes, in a gradually improving international climate, is an important aspect of present United States foreign policy. It is extremely difficult to make a political decision deviating sharply from the course of traditional commitments, and such a ''chain reaction'' of small steps, each encouraging the next, may indeed hold the greatest hope of breaking away from the ominous threat of a perpetual arms race.

A freeze on the levels of nuclear armaments is the next step being pursued most vigorously at present as a part of United States policy. If successfully negotiated, it may come in the nick of time to prevent the initiation of a long and complex new round of the arms race, with missiles intended for defense calling for ever-increasing numbers of missiles for attack. Thus, it is very important that we try to negotiate small steps, including specific measures of arms control, if we can.

The trouble with relying on successive negotiation of many small steps is that the limited advantages inherent in some of the small steps may not provide enough incentive to overcome the obstacles to agreement. If one hurdle is high, the horse may not jump it unless he knows that there is hay in the barn beyond a few more hurdles.

Some small steps toward disarmament are possible with little or no inspection. The first high hurdle on the way to disarmament comes when the number of nuclear weapons is reduced to such a low level that detailed inspection of large areas is needed. Far-reaching inspection is needed at low arms levels in a cautious approach to disarmament because without it there is a possibility that enough nuclear missiles might be hidden to alter seriously the balance of power.

Area inspection is very hard to negotiate in successive small steps. You either decide to let inspectors roam around the country, or you do not. The Soviet Union tries to compensate for its smaller number of intercontinental missiles by maintaining greater secrecy as to their locations than we are able to do in our open society. We can only hope to induce the Soviets to relinquish their secrecy if we are willing to provide an incentive to get over this hurdle in the form of a big step toward disarmament—a big step bringing real disarmament in sight at the end of the course. This requires negotiating more than one small step at a time.

Inspectors with free access to a whole country are apt to uncover missile locations almost anywhere. This fact, that "inspection and secrecy don't mix," seemed to be a definite technical roadblock in the way of disarmament, rather than just a hurdle, until a few years ago when it was suggested that both inspection and disarmament should be introduced into one region after another of each country involved.[1] This means that there would be a sequence of small steps, but all planned in advance and negotiated in one farsighted agreement. The regional idea sounds rather obvious and elementary when now described, but the fact that this possibility was overlooked during many years of disarmament debate suggests both that it was the result of considerable thought and that it is always worthwhile to go on seeking good new ideas about how to move toward a safe and disarmed world.[2]

A crucial problem in orderly disarmament is to know *how many* weapons there are, without needing to know *where* they are until their turn to be disarmed comes. The "region-by-region disarmament plan" includes a rather special way to use the stepwise disarmament procedure as a sampling process to build confidence that each side is telling the truth about how many weapons it actually has to be disarmed. At the same time it avoids letting other countries know *where* the weapons are, which might tempt some country to try to destroy almost all of them in a surprise attack, and thus to minimize reprisal.

Region-by-Region Disarmament

Let us see in outline how the region-by-region procedure would work. First, an international disarmament authority must be set up and must organize a large group of inspectors. Each country divides its area into a number of regions, let us say six regions of approximately equal military value, and then makes a declaration listing the numbers of weapons and other military objects of various kinds in each region. Since nations seem to suspect one another, there may be grave doubts at first whether the lists are honest. It is agreed that on a certain date there will be an unpredictable selection of one region of each country to be sealed off and disarmed in the first stage. This might be done "by lot," or the "other side" might be permitted to choose.

When the choice is made, the inspectors quickly concentrate on the borders and transportation facilities of the selected regions to prevent shipment of arms. Each nation is then required to submit a detailed list of where the weapons previously reported for that region are to be found. The inspectors observe them and see that they are dismantled, if necessary in such a way as not to learn certain critical details of their construction. At the same time and perhaps for the rest of a period of about a year, the inspectors are free to roam through this region and assure themselves that there exist no significant quantities of hidden arms (particularly missiles and aircraft). Once they have done this, and found that the number of weapons agrees with the number on the original list, there is reason to begin to believe by this process of "random sampling" that the original list is honest.

After about a year, the process is repeated in a second region, and so on. At each step the basis for confidence in the original list is increased. Toward the end of the process, this basis for confidence could be very important if, at the last stage, each nation is required to relinquish the last of its nuclear arms in the sixth region and would want to be sure that others were doing the same. Another important feature is that, until a certain region is chosen to be disarmed, the secrecy of location of missile sites and other installations in it is retained.

This is the bare outline, and several modifications and special features could be discussed. For example, there is the problem of how missile-firing submarines and their bases are to be included, perhaps as a separate "region," and there is the possibility that certain types of arms can be retained in an otherwise disarmed region for the sake of local protection.

A more important problem concerns the kind of world stability that can be planned for the end of the process of national disarmament down to the local police level. While all matters of local management and all aspects of national sovereignty, aside from the means of making war, will presumably be retained by national governments, it appears to be necessary to establish some form of reliable world military force capable of preventing any nation from rearming and an equally reliable world political body to make decisions about the role of the world military force. This poses extremely difficult problems, but should not be permitted to prevent the active search for substantial disarmament soon.

The "Transitional Deterrent"

Perhaps the most hopeful possibility is that both sides will recognize that it is too difficult to agree in advance on the details of the final solution until we have had more experience living together with lower, and carefully controlled, levels of armament. This leads to the idea of the "transitional deterrent": a carefully balanced and stabilized low-level nuclear deterrent force to be considered as a transitional stage on the way to complete disarmament. That each side could reasonably go to lower levels under appropriate conditions follows from the fact that nuclear weapons are individually so destructive that neither side needs all of its present supply to inflict on the other side more damage than the other side can rationally consider "acceptable." A single nuclear weapon can carry more explosive power than was used in all of the wars of history, including World War II (and a single long-range missile of the usual type almost that much). Our plans are to have almost two thousand long-range nuclear missiles by about the end of this year. One hundred would constitute a stupendous threat, if the other side had no more. This lower level of missiles would be just as effective as would a threat by uncounted thousands, in providing a "nuclear umbrella" under which East-West struggles over various national boundaries could proceed until such problems are resolved.

Thus, a reasonable "transitional deterrent" might be composed of one hundred missiles (or two hundred, or fifty) on each side, reliably counted and installed in such a way as to be invulnerable to attack from the other side, invulnerable in the sense that several missiles on the average would be required to destroy one, so that no attack by an approximately equal number could destroy most of them and prevent retaliation. Missiles are protected in holes in the ground called "silos." One way to ensure the security of the remaining missiles would be to have three times as many silos as missiles. Inspectors could then inspect groups of a dozen silos at once, counting the missiles in them, after which the missiles might be shuffled among these silos without inspection. Another way would be to use submarines as missile bases.

The region-by-region disarmament procedure can be employed as a reasonable and cautious way to set up a transitional deterrent. As each region is opened to inspection, the weapons in it could be destroyed with the exception of the silos, leaving an appropriate fraction of the missiles as part of the transitional deterrent.

This, then, provides the possibility for an agreement on a large package of disarmament at once, one which should have greater mutual advantage to the two sides than does any small step in arms control, without requiring a final commitment to go all the way to a disarmed world. That would be left for future negotiation. Its advantages are that it provides both an alternative to a perpetual arms race, one that may prove negotiable, and that it constitutes a large step toward a situation much more favorable for the further design and negotiation of the conditions for a disarmed world. For instance, there will be much greater certainty of the numbers of weapons that remain to be disarmed.

In view of the fantastic dangers of a limitless arms race among an increasing number of nuclear nations, this would seem to be a far safer course for our nation to pursue as the primary goal of its foreign policy. If we are to pursue it effectively, we will have to make the difficult decision to accept military parity with the Soviet bloc, rather than to try to remain well ahead. This does not necessarily mean exactly equal numbers of long-range missiles, but it might mean this if other items of military strength are to be pared down accordingly, after due allowance for geographic differences.

It may, with a grain of truth, be claimed that we have already made the proposal of a region-by-region approach and of a transitional deterrent, and that the Soviets have failed to accept it. In an outline[3] first submitted in the negotiations at Geneva on April 18, 1962, we proposed in a sketchy way three stages leading at the end to general and complete disarmament and therein *suggested,* without including it as an actual proposal, that the aim of establishing inspection in proportion to the amount of disarmament at each stage (Section G3c) "might be accomplished, for example, by an arrangement embodying such features as" the region-by-region procedure described above. The suggestion was not integrated with the transitional deterrent and the end stage of complete disarmament was alluded to so vaguely as to leave the suggestion no more than a trial balloon for Soviet reaction, which has not been favorable. Elsewhere in the negotiations it was made clear that we had in mind proportionate reductions in nuclear strength, to leave us a fixed numerical ratio of advantage over the Soviets as the numbers of missiles became smaller, rather than an approach to parity.

We have negotiated separately for the establishment of a lower-level deterrent, rather than working toward general and complete disarmament as the immediate goal, on the grounds that without some nuclear deterrent a few hidden nuclear weapons could be decisive. It is a hopeful sign of progress in the slow negotiations that the Soviets have finally acceded to this concept in principle—at least since Gromyko's United Nations speech of September, 1962. They seek, however, to negotiate the final stage of complete disarmament in the same package.

A promising line of approach to the disarmament problem would be for us to propose a treaty providing in realistic detail region-by-region steps to a stabilized low-level transitional deterrent, incorporating a rapid approach to parity in both nuclear and conventional weapons (which means destroying more Soviet heavy tanks and short-range missiles and more of our long-range missiles as a modification of the first regional stages). This would be proposed with the sincere intention of using the breathing spell provided by the transitional deterrent for the development of international machinery for the maintenance of a stable peace in a world without national arms. It would be to the mutual advantage of the two nuclear giants, without favoring one over the other, and to the other countries of the world as well, because it would make the scourge of nuclear war less likely.

It would make war less likely, among other reasons, first, because there would be fewer weapons and weapons commanders to fire by mistake; second, there would be less pressure, on the part of national leaders under some provocation, to attack to exploit a supposed temporary advantage in rapid weapons development, before the other side advanced still further. Another mutual advantage was implied by President Eisenhower when he said, "Every gun that is made, every warship launched, every rocket fired signifies, in the final sense, a theft from those who hunger and are not fed, those who are cold and are not clothed."

A proposal of this type is more clearly to the mutual advantage of both sides than previous disarmament proposals, each of which has somehow been slanted in favor of the side proposing it. For this reason, there is a greater chance that the Soviets would accept this than our earlier proposals. Whether they would actually accept it or not cannot be known without trying.

NOTES

1. See Louis B. Sohn, "Disarmament and Arms Control by Territories," *Bulletin of the Atomic Scientists,* April 1961.
2. See D. R. Inglis, "Region-by-Region Disarmament," *New Republic,* June 27, 1961, and "Disarmament after Cuba," *Bulletin of the Atomic Scientists,* January 1963.

3. Outline of Basic Provisions of a Treaty on General and Complete Disarmament in a Peaceful World; U.S. Arms Control and Disarmament Agency, Publication 4, General Series 3, May 1962 (Washington, D.C.: U.S. Government Printing Office, 1962).

Evolving Patterns of Nuclear Disarmament Proposals

[1962]

To most of those participating in the historic test of the world's first atomic bomb at Alamagordo, N.M., in 1945, it was evident that world civilization could continue to evolve in a fruitful way only on the basis of innovation in international relations as radical as the technical innovation which had unleashed the new atomic destructive power. These scientists were not so naive politically as to believe that this would be easy, for political developments are clearly not so nicely subject to abstract formulation as the affairs of the laboratory and proving ground. Yet the technical difficulties and the human obstacles of organization and priorities which preceded that cataclysmic flash above the New Mexico desert were so great as to lead to the hope that men might mount efforts of a similar magnitude and ingenuity to overcome the even greater political difficulties in the way of worldwide atomic control. The basic problem was, and still is, that a factor of a thousand or a million in the destructive energy of the largest weapons in the hands of man means that another general war would be an unthinkable tragedy, enormously more destructive than the very destructive wars of this century; yet this tragedy is very likely to take place because there always have been wars and, without great political development, apparently always will be, as long as men and competing nations exist.

The shining ray of hope beneath the dark clouds of the nuclear age emerges from the possibility that the very factor of a million which so magnifies the prospective horror of nuclear war will somehow act as the new element in the situation to bring an end to large-scale wars, and indeed to all wars. The threat of retaliation to an aggressor is so much more powerful than ever before that one is tempted to hope that a system of mutual deterrence can be developed to discourage aggression for a long time. Many people find assurance in the fact that it has already worked for the first few years of the increasingly dangerous threat, as a new and paradoxical mixture of delicacy and insensitivity in international behavior has appeared. Yet with the prospect of many nuclear nations of the future, each threatening the other with nuclear retaliation, and with the possibility of the accidental explosion of this tense powder keg, it seems naively optimistic to hope that this uneasy peace by reciprocal threat of ever-expanding nuclear reprisal can last for many decades. But this is a temporary expedient with which we now find ourselves intensely occupied almost to the exclusion of serious national consideration of any other course. It cannot but lead to disaster if it is not replaced by some more satisfactory permanent solution to the problem of world stability. Our having lived through a span of the nuclear age about as long as the span of time between world wars I and II gives us only a false sense of comfort but does not prove a thing.

The alternative to an ever-expanding and increasingly dangerous arms race is some form of limitation of arms and of the development of arms. The cherished concept of national sovereignty and the consequent military tradition of seeking as much power as possible, with which to be ready to impose the will of one nation on another, are immediately encountered as almost insuperable political obstacles to any serious thought of arms control or disarmament, yet this very military tradition can no longer provide us the security basis for long-term national and civilized existence. The political obstacle is aggravated by a primary technical obstacle; as long as we depend on military might for deterrence, even temporarily, some degree of military secrecy is considered essential (and is achieved more successfully in the East than in the West). Yet control of arms requires international knowledge of existing arms which interferes with military secrecy. For this and other reasons, the military authorities in various countries tend to oppose arms control and disarmament, unless and until they themselves can become convinced of the long-range futility of their profession in the face of the nuclear threat, as some of them already have. And for this reason also it is necessary to consider various schemes and levels of arms control and disarmament and to promote most seriously those which are least incompatible with the maintenance of some degree of military secrecy, or at least those which place no nation at a distinct disadvantage in this respect, as long as world stability remains dependent on a military balance. We must consider both arms control and disarmament and try to be ready to use whatever degree of one or the other that best serves our purpose of long-range survival and growth and may at the same time be acceptable to other nations.

The first, and to date the most serious, attempt to control nuclear arms was made in the first year of the atomic age, under the name of the "Baruch plan," more properly called the Acheson-Lilienthal plan because Lilienthal, at least, had more to do with its imaginative origin. It was an attempt to nip the nuclear arms race in the bud before it got started. In it the United States proposed to relinquish its new monopoly on primitive atomic weapons if all nations would grant a monopoly in nuclear weapons research to an international authority whose purpose it was to assure that no nation would develop a nuclear military capability. It failed, probably because the world was too slow to wake up to the need. Our nation was enlightened enough to make the proposal in good faith, and, as an important step in the negotiation, this plan could have served as the cornerstone for a marvelous new age if it had been received with equal enlightenment. Yet we ourselves were so hesitant about it as to put the negotiation in the hands of our conservative elder statesman whose words rang more with caution than conciliation, and the Soviet Union was then ruled by the iron hand of Stalin and would have none of it. It was before its time, yet it could have been proposed at no other time. *Now we must make do with less, and with less and less as time passes.*

It is very important that we thoroughly appreciate this fact as we waver between disarmament and the arms race, as we consider how far we can afford to go at this stage in compromising military desires for the sake of turning toward the disarmament goal. We must intensely feel how time is fleeting, how inexorably the refinement and proliferation of nuclear destructive capabilities are making the technical problem of arms limitation more difficult, and how rapidly the prospects are shrinking for accomplishing the initial acts of restraint on the road to disarmament.

In trying to develop an appreciation of the changing nature of disarmament opportunities, it should be helpful to have available a complete study from that earlier period with which to compare our present-day disarmament problems. One such study from ten years ago, which culminated in a rather specific proposal, is published here for the first time [concurrently in 1962 and in abbreviated form as the first article of this chapter], partly in the hope that it may be useful in this way as a guidepost or point of comparison with the past. This study, entitled "A Specific Proposal for Balanced Disarmament and Atomic Control," dated December 1, 1952, was not published at that time because it was felt that it might be more useful if privately circulated in appropriate governmental quarters, more useful in particular, very optimistically, as a possible point of departure for negotiation.

It will be seen that by then, in the seventh year of the atomic age, the central problem was no longer the prevention of independent national production of and research with fissionable materials, as it was in the Baruch plan. Instead, two national stockpiles of these materials had already been produced and it was hoped to estimate the extent of past production accurately enough to be able to retire most of the stockpiles of these dangerous materials with reasonable confidence, in spite of the inability of detection methods to find a hidden cache of them. The principal impasse in negotiation, however, arose from the great disparity in the types of forces in which the two sides excelled. The United States was vastly superior in nuclear striking power, but the USSR was conceded a distinct military advantage in conventional forces because of greater military manpower and more direct lines of communication within a compact continental land mass. Perhaps the principal innovation of the old proposal here published is its meticulous approach to the problem of balancing concessions made by the two sides in quite different categories of weapons, in order to provide a proper *quid pro quo* despite the technical asymmetry. Similar problems may be faced in arranging the details of a future application of the modern region-by-region disarmament plan which is discussed in this article.

We four authors of the "Specific Proposal" had all worked on the development of the first atomic bomb during the war, and with this background our primary emphasis was on finding a way to bring the stockpiles of nuclear weapons and materials under control. The secondary aspect of the problem was to match atomic concessions with steps in bringing other military preparations under control at the same time, so as not to be unfair to either side. The study was made before the appearance of the H-bomb on the scene, although we knew it was coming, and the proposal was completed and circulated just after the first thermonuclear test. We had hoped during our study that there would be time to promote negotiations along those lines before the actual advent of the H-bomb.

The successful testing and subsequent rapid development of the H-bomb technique greatly increased the significance, in potential destructive power, of a possible small hidden cache of fissionable material. This made it more important than ever that primary emphasis be placed on the nuclear side of disarmament. It made it more important than ever to have accurate estimates of past production if control of nuclear materials could still be attempted at all, and, if not, to put much greater emphasis on the control of the means of delivery of nuclear weapons.

The writing of the "Specific Proposal" left its authors with the firm conviction that an adequate exploration of the possible ways of extracting increased security from arms limitation agreement was much too vast and vital an undertaking to be left to the initiative of small private groups. The need for greater national effort in this direction became acutely clear as negotiations proceeded on the basis of inadequate preparation. By grappling with the problem in some detail one could appreciate the magnitude of the task, but those who had not seriously grappled with it were not easily convinced, or found it hard to appreciate, that progress on the technical side could be helpful even before the political desire for progress arose, perhaps as a catalyst for that desire.

In the years 1952–57 the disarmament negotiations pursued a new course in which combined atomic and conventional disarmament was discussed. However, in the light of the considerations of our "Specific Proposal," it seemed that the official discussions never got to the point of being realistic about considering the nuclear control side of the problem in practicable detail. The primary emphasis was on the quotas for military manpower, in various proposals varying from one to two and one-half million men each for the United States, Russia, and perhaps China, with about three-quarters of a million each for Britain and France. The proposals contained various vague references to "cutoff of new production of fissionable materials" and to studies by a group of experts to design a control system, but they always seemed ill prepared and preliminary, as though it were too much trouble to think about these technical matters seriously. It is perhaps notable in retrospect that the Russians took the lead in proposing in March of 1957 the abolition under international control of all military missiles.

Throughout all those negotiations, the world public was sadly deceived by being given the impression that real negotiations were going on and that their failure was proving disarmament to be unattainable. The truth seems to have been that it was not being sought in any determined and consistent way by any great nation, though there were surely some dedicated individual negotiators whose only fault was that they were not backed up by national policy. There was an aura of unreality about the negotiations which reflected a lack of agreement among officials back home on what should be negotiated, if anything. Whereas disarmament demands deeply studied preparation, bold decision, and decisive steps, delegations were sent informed only by a few weeks of last-minute briefing and study, backed by complete indecision as to whether disarmament was seriously desired, and able to propose only mincing steps. One serious roadblock was the unwillingness of the Soviets to be definite about what they meant by inspection, and much was made of this in the Western press, but we made no proposals sufficiently definite to explore their willingness adequately. Most proposals seemed to be made for their propaganda effect, in order to appear to the world to be sincere while actually being so favorable to the proposer as to have almost no chance of being accepted. When this chance was misguessed and the other side surprisingly accepted, the proposal was withdrawn.

Two examples of such withdrawal, when our country showed quite definitely its lack of decision to seek disarmament, occurred in 1955 and 1957. (A third occurred in connection with avoiding the possibility of a test ban agreement in late 1958, as will be discussed further here.) The first example was perhaps the most clear-cut one. In April, 1955, we were chiding the Russians for insincerity about disarmament in not accepting our very reasonable (if sketchy) proposal. Then on May 10, 1955, the Russians suddenly accepted essentially the entire proposal. There were still details to be cleared up, but the acceptance was so nearly complete as to be a great breakthrough in Russian declared willingness to accept inspection, so great indeed that if we had had any desire for controlled disarmament, we would have eagerly pressed the negotiation further to clear up the final points. Instead, we simply broke off the negotiations and on September 6, 1955, withdrew our proposals. Our failure was even more serious than that, for the Soviets at that time threw in an extra proposal which represented a very substantial parting of the Iron Curtain and would have been really valuable to us as a starting point for arms control inspection. They proposed the establishment of ground ''control posts'' at ports, railway junctions, on main roads, and at airports to help guard against the danger of surprise attack. The really surprising thing is that we did not take it as offered, but instead said we would only take it if they would give us ''open sky'' flyover inspection in addition. This failure is a measure of our abject unpreparedness for meaningful negotiation.

The 1957 incident was the one in which the progress toward arms limitation and a cessation of nuclear tests being made at London with Governor Stassen as our eager representative apparently seemed too dangerously near success to the then chairman of the AEC and his Livermore advisers who suddenly unveiled the ''clean H-bomb'' as an object of our atomic development so urgent as to lead President Eisenhower suddenly to withdraw our representative from the talks. This precipitate decision, like many others, was notable for its emphasis on what we would like to have in our arsenal, and was made with utter disregard of what others will develop in theirs if we fail to reach a restraining agreement.

The ''Specific Proposal'' was an attempt to show how a proposal could be balanced so as to be fair to both sides if the political desire should arise to reach a disarmament agreement. The proposals which instead reached the light of day showed that such political desire did not exist. The tragedy of it was that precious time was being lost to introduce some moderation, as the further development of nuclear weapons promised to make them rapidly more dangerous and more difficult to control. . . .

Now that we have been lucky enough to live with the H-bomb for ten years and stockpiles have been accumulating apace, there is no longer any practicable hope of retiring the stocks of nuclear explosive materials in the way outlined a decade ago in the ''Specific Proposal.'' The big national stockpiles are now measured in the tens of thousands of megatons. Even if these were so thoroughly eliminated that there would remain some uncertainty about only the last 10 percent (and the specific proposal estimated a 20 percent uncertainty then when the job was easier), this 10 percent could be enough to kill more than half of the population of a great country. The hope now is to control means of delivery rather than the warheads themselves and, at the same time, to move toward a political situation in which hidden stockpiles of nuclear bomb materials alone will be irrelevant. One approach to such a political situation has been studied by Clark and Sohn in their book *World Peace Through World Law.* . . .

Complete disarmament, if it were politically acceptable, would have one unique technical advantage. With complete disarmament, neither side would have military secrets to guard, and thus there would be no military reason for not letting inspectors roam at will. Even Premier Khrushchev, who has seemed allergic to inspectors when partial disarmament was discussed, has emphasized that the USSR could afford to admit inspectors freely to verify compliance with a plan for complete disarmament. This is a very simple and logical deduction, and we should consider it fortunate that the Soviet leaders recognize it. However, standing alone, it is not very useful, until we can find a way to get from here to there without taking too much of a risk along the way.

Recently there has been suggested a way to circumvent the difficulty that inspection and military secrets do not mix, a way to take advantage of the Soviet leaders' view that inspectors should be acceptable when (and thus presumably where) there are no military secrets. The plan, as proposed by Professor Louis Sohn of Harvard, is so simple as to make one incredulous at first that it could have any unique significance. On appreciating its revolutionary efficacy, one wonders why it was not invented long ago. That it was not is a sorry indication of the inadequacy of official studies. If there had been established a National Arms Control and Disarmament Agency on an imaginative scale a decade ago, this simple idea might have been available to bring American disarmament policy into fruitful focus during the formative years of Soviet policy following Stalin's death. The present Soviet avowed interest in total and complete disarmament (in four years!) may indicate that it is not yet too late to work out a reasonable compromise, passing in a very few years to a low-level transitional deterrent, with complete national disarmament under an international force as the real ultimate goal. . . . [An explanation of the plan appears in the previous paper and is here omitted.]

This region-by-region technique is an important innovation in the disarmament field because it circumvents what appeared to be a technical impasse. Without it, great trust is needed at some point, perhaps trust that, in spite of the inadequacy of minimal inspection designed not to interfere with military secrecy, the other side will not try to evade the conditions of a disarmament agreement. With it, nations very suspicious of one another's intent to evade where evasion is possible can still reasonably enter into an agreement in which growth of confidence is based on thorough inspection to verify faithful performance. If two or more very suspicious nations recognize the great mutual benefit of genuine disarmament, this plan makes the difference between technical possibility and impossibility of exploiting the mutual interest.

The region-by-region plan is, however, only a means to pass safely from one level of armament to another. Ideally, it could take us right down to complete and universal disarmament if we were to be ready for it by the time the last region would be disarmed, say, in seven years. It is to be assumed that we will not be ready so soon. It will probably take much longer than that to resolve the ground rules of the competition between East and West over the "uncommitted nations" to a sufficient degree that trust can be generated in a decision-and-command apparatus for an international police force. For this reason, the region-by-region plan should be used to reduce the level of strategic deterrence to the lowest and least accident-prone level that can be agreed upon to maintain an interim military balance, or transitional deterrent.

The purpose of establishing a transitional deterrent stage is thus to make it possible to get started on a very substantial degree of disarmament now, without waiting to solve all the other problems of the troubled world first. The important fact here is that the conduct of limited power politics in local situations, where both sides intend to stop short of triggering a general war, is independent of the exact size of the strategic deterrent. The strategic nuclear forces which both sides are planning to build up in the next few years are to be far in excess of what will be needed to inflict unacceptable damage and to promote restraint where national interests conflict. A force of the order of fifty missiles on each side should suffice as a threat of unacceptable damage.

There are doubtless several ways in which such a force could be arranged so as to be inspected and yet retain sufficient uncertainty of location as to be invulnerable. One way is to have the fifty missiles deployed in an unknown way among two hundred hard sites, or "silos," of which some hide real missiles and some are empty or hold decoys. They could be subject to occasional inspection on a regional basis, and subject to secret reshuffling among the silos within each region in such a way that the inspectors would know the exact deployment of only a few of them at any one time, and yet after covering all regions, they would know the exact number of missiles.

One of the great advantages of the region-by-region plan over the old "Specific Proposal for Balanced Disarmament" is that the former does not require the negotiators to agree in advance on a scale of relative values of various types of military hardware. Both plans call for a number of stages, seven of them for example, but arrange to balance the concessions of the two sides at each stage in quite different ways. The region-by-region plan puts the burden of making an equitable division on each country for itself, and calls on the self-interest of each country to make the stages proportionate on the two sides. A country participating in the region-by-region disarmament procedure would presumably not want to divide itself into regions of greatly different military value, for fear of being left to depend for deterrence on the weakest region at the last stage.

The Baruch plan, the "Specific Proposal for Balanced Disarmament," and the region-by-region plan represent three partial answers to the nuclear disarmament problem at three stages of history. The types of control they envisage are successively less tight in response to the continuously deteriorating prospects and emphasize the urgency that a start be made on arms limitation soon. Each of the three plans proposes that we tackle the problem in a drastic fashion by taking a bold decision to change the course of the arms race at once. Each of them was motivated by an adequate recognition of the danger of our present drift with the flooding current of arms proliferation. In contrast with these forthright plans, which would have run contrary to the spirit of freedom of enterprise to destroy ourselves, there are being discussed these days various gradualist approaches which would not so severely tread on the toes of vested interests, and are perhaps looked upon with more favor in influential circles. These gradualist approaches include mincing steps in arms control, some of them not very different from the proposals of the mid-fifties to reduce the danger of surprise attack, such as "flyover inspection," some of them going a bit further and placing emphasis on the possibility of arms self-control, by the two sides, to put emphasis on second-strike capabilities.

While any tendency toward contagious self-control is surely to be encouraged, the trouble is that these gradualist approaches are apt to buy too little, to permit the proliferation of weapons in refinement and in the number of countries having them, and to do no more than reduce the likelihood of war per year without providing any long-term solution of the nuclear age instability. The advantage of some of them is that they require no formal agreement, and so might be politically practicable at an earlier date than a far-reaching disarmament agreement. Such steps are fine if they are not considered to be a substitute for disarmament agreement, but rather a prelude. Others of them do require formal agreement on mincing steps of arms limitation, but steps disproportionately small in relation to the amount of inspection they propose. They do not offer to pay, in disarmament advantages, for the cost in terms of inspection and are therefore completely unrealistic as a basis for agreement. It is important that plans of this sort sometimes proffered for propaganda effect be recognized as such and that their failure should not be permitted to continue to discourage the body politic about the prospects of disarmament agreement. Insofar as they are not genuinely nuclear disarmament proposals, their details fall outside the scope of this discussion.

The study of the three successive disarmament plans, the Baruch plan of the mid-forties, the "Specific Proposal for Balanced Disarmament" of the early fifties, and the region-by-region plan of the early sixties, should show that the difficulty of arriving at a serious disarmament agreement is not the lack of reasonable technical means to accomplish the disarmament goal without jeopardizing the vital interests of either side. The difficulty seems rather to be the habit of continuing with traditional, prenuclear attitudes toward the military solution of international problems and the consequent lack of a mutual will to agree to disarm down to controlled low levels, as a serious answer to the very real nuclear threat.

6 Attitudes and Decisions

In our democracy as it has evolved within the framework devised by the founding fathers for much simpler times, the decision-making process has become largely a contention between competing influences on Congress and the executive branch. In developing the perquisites of peaceful prosperity, and even some of the arts of war, healthy commercial competition has been accompanied by competition in influencing the government for special treatment. The philosophy "What is good for General Motors is good for the country" has served well. With half a continent to be exploited by free enterprise, the country has prospered marvelously despite duplication of effort, waste, and uneven distribution of rewards.

In this last half-century we face an entirely new situation. In a world loaded with weapons of utter destruction we are faced with a difficult adversary sharing with us at least a common interest in survival. One wonders if our traditional democratic process of making decisions by competing pressures is up to the task of charting a safe course through these dangerous waters. It is a new and crucial challenge.

In a well-regulated nation, civilian authorities, the statesmen, should determine the level of military preparedness required by present circumstances, and the function of the military should be efficiently to provide that level of preparedness. The statesmen, not the military, should be concerned with influencing, through arms control agreements, the military level mounted by potential adversaries. The function of the military establishment is such that military personnel quite naturally want more and better weapons. It is natural that they and their suppliers should want to influence the statesmen, but this influence should not preclude mature judgment. Up until World War II, we had a relatively reasonable balance in this respect. The oceans provided confidence that our land would not be attacked, and the level of weaponry maintained, perhaps aside from battleships, was fairly modest.

The end of World War II seems to have changed all that. It left an ideological split between the victors who had become strongly armed adversaries and it left the world with the atomic bomb that nullified the protection of broad oceans. The military establishment, strengthened in wartime, retained its preponderant influence after the war, and never returned to its former modest peacetime status.

With the impact of costly mass media now available the big institutions with the largest budgets have the most influence. It is a matter of "those who have, get." Perpetuation of established ways becomes the norm; striking out on a new course is difficult. Vested interest in current expenditures exists at the level of individual jobs as well as at the corporate level. Politically, a bird in the hand, or a defense contract job in hand, is worth two hypothetical jobs, after conversion to civilian production, in the bush. The attitudes of large sectors of the public are shaped accordingly. All through the atomic age, as we see repeatedly in these papers, the hawks have had an advantage over the doves in this respect.

The problems of the nuclear balance have seemed too arcane to many voters who place a simple trust in military judgments. Elections have turned more on pocketbook issues than on matters of statecraft, and bloated defense appropriations have ground on uninterrupted by thoughts of serious negotiation. Elected representatives under conflicting pressures find it difficult to think as statesmen. Officials charged with negotiations, notably Ambassador Stassen in the early fifties and, later, most of those in the Arms Control and Disarmament Agency, have worked hard to make progress toward arms limitation agreements with the Soviets, only to have support withdrawn when success seemed near. That story is told in the 1974 "Aspirations and Frustrations" article toward the end of this chapter.

In 1955 I wrote "We Haven't Really Tried." Those who conceived the Acheson-Lilienthal-Baruch plan had really tried. President Truman and his secretary of state, Dean Acheson, in adopting this proposal, had made an attempt to head off the nuclear arms race, but in selecting elder statesman Bernard Baruch, with his caveats of conservative caution, to present the proposal, they hadn't really tried. In 1955 the complaint was that the administration had tried nothing else when this proposal failed of acceptance.

Now at this late date we can well ask again, "Have we really tried?" There is no question that many devoted individuals both in and out of government have tried very hard and patiently, but they have been permitted to try only within the limitations of a trend of national policy that values protecting our latest additions to overkill over attaining the benefits of mutual arms reductions. In this sense, we still haven't really tried. We have at no point given arms control the priority it deserves over continued nuclear arms production in a quest for improved long-term national security and indeed for the prospect of human survival.

Will we ever really try and succeed? Whether or not we take a genuine initiative, realistic enough to elicit a favorable Soviet response, may depend largely on the world view of an American president and his ability to discount Pentagon briefings. Public attitudes toward the arms race have been divided. The entrance of a white knight on horseback, so to speak, may be needed to resolve the issue. President Kennedy understood the problem and it was once hoped that he, after paying his respects to a rapid buildup and after experiencing the Cuban trauma, might be the one. We may still wonder what significance an assassin's bullet may have had for the fate of the world.

The next three presidents in the decades of the sixties and seventies each took some initiative in arms control but none put arms reduction ahead of arms procurement. In almost continuous negotiations we have tried, but with limited objectives. Each step of agreement considered by the Arms Control and Disarmament Agency had to be approved by the Pentagon. The negotiations have progressed much too slowly to keep up with the mad pace of technological advance. The limited successes achieved have amounted to little more than slightly slowing the arms race while legitimizing each next step of the buildup. The prevailing mood in Washington seemed to tolerate only negotiations aimed to preserve a position of superior strength, with little appreciation that superiority has no real meaning when the overkill capacity of both sides is so enormous. All along the negotiations served partly as a smoke screen to obscure the steady growth of stockpiles and partly as justification for new weapons systems needed as bargaining chips. We have acquired a formidable arsenal of bargaining chips.

The term "arms control" has two rather distinct meanings. At first it was usually used to mean controlled partial disarmament as distinct from complete disarmament, with emphasis on the mutual advantage of achieving greater safety at a lower level. It may also mean control of the balance in a continuing arms race or limitations at a steady high level reached by competing attempts to gain unilateral advantage through hard bargaining. In most political discussions the latter usage unfortunately seems to prevail.

The cautious negotiations did accomplish something, but not nearly as much as they might have if given broader scope and higher priority. The partial test ban treaty was achieved in 1963 but permitted continued weapons testing underground. Negotiations toward the rest of the test ban since then slowly approached a formulation acceptable to both sides but did not stop testing. The anti-ABM treaty incorporated in SALT I, by avoiding the destabilizing ambiguities about the effectiveness of a dubious shield, has doubtless avoided demands on both sides for even greater numbers of attack missiles. The SALT II negotiations leading to an unratified treaty did at least probe the limits of the willingness of both sides to exercise restraint in future weapons planning.

With the arrival of the new administration in the new decade, we ceased to make any realistic try at all. All this cautious striving to work out useful agreements with the Soviets was peremptorily abandoned. The test ban negotiations were cut off. The SALT II treaty that had required seven years to find common ground acceptable to both sides was left unratified, though its limited provisions continued to be observed. Instead of carrying on from there, the negotiations, renamed START for Strategic Arms Reduction Talks, served as a forum for starting all over with demands more favorable to our side. Our demands included insistence on establishing an intermediate-range missile balance on land in Europe without counting French and British forces or our submarines, and also insisting on discussing intermediate-range and strategic balances in separate negotiations with no trade-offs between the two.

Specific proposals submitted in the negotiations gave the illusion of a favorable attitude toward arms reduction but were in reality so biased in our favor as to be clearly unacceptable to the Soviets. It is important to appreciate how deceptive appearances can be in this way. The "build down" proposal is a good example. If successfully negotiated it would actually reduce the number of missiles and suggests at least that some officials appreciate that we have more missiles than we need, while wanting to modernize our arsenal. The proposal was for each side to retire two old ICBMs for each new one deployed. The catch is that the obsolescent missiles to be retired hark from the early sixties when we had about four times as many of them as the Soviets had, so we would be deploying proportionately more modern ones. It was a scheme for regaining a lost superiority, but it contains the germ of an idea for politically feasible initial arms reductions if the bias can be negotiated out of it. That our terms were obviously unacceptable to the Soviets was emphasized by, among others, former secretary of state General Alexander Haig, who favored realistic negotiations but who was replaced early in President Reagan's first term.

As if to further guarantee that the negotiations would be no more than window dressing, the president's style was to madden the Soviet leaders and at home embellish his popular image of genial toughness by vilifying the

Soviets vehemently and joking undiplomatically about his power to annihilate them and then, shortly before election time, to seek to "bring them back to the table." Never before had the smoke screen function of negotiations as an end in themselves been quite so apparent. Then, when the negotiations were resumed in his second term, the president's words were more moderate as he called for deep arms reductions. However, his insistence on pursuing his dramatic "Star Wars" dream of making nuclear missiles obsolete, projecting the arms race needlessly into space despite Soviet protests, seemed to be the new assurance that the negotiations would not interfere with the steady progress of the arms race.

That President Reagan with his background and prejudices should act thus is perhaps not surprising. It is frightening that such a large proportion of the American people go for it. That he could so successfully obscure his blunders and hawkish policies with simplistic dovish talk and confident hopes for technical miracles and still win reelection brings into question anew whether the American electorate can ever meet its responsibility to save the world from the dire consequences of an endless arms race. With all our heterogeneity, too few of us seem to feel the responsibilities of citizenship in this regard. Too many let themselves be misled to believe that, with all our awesome nuclear strength, we need to be still stronger and that spending a lot of money on any weapons system, no matter how destabilizing, makes us stronger.

The responsibility is indeed ours. We have all along been the leaders of the race. The decision is ours (or at least has been ours) if we collectively choose to make it, rather than to let it be made in default by attitudes in Washington perpetuated by that enduring triangle, the military-industrial-political complex. We are the only electorate in the world with the power to stop the race. There is no meaningful Soviet electorate.

There is a new spirit in the land and indeed in the Western world, a new upwelling of public opposition to the arms race. It gives hope, even in the face of adversity, but was not strong enough to deny the president a second term. There is a new political constituency for doves, for politicians favoring arms moderation through negotiation. It includes the freeze movement seeking to stop the nuclear part of the arms race in its tracks before shifting into reverse. It is an outgrowth of decades of educational efforts by many organizations but was powerfully spurred in reaction to President Reagan's blunt and enthusiastic disclosure of Pentagon intentions and attitudes.

Almost everyone has to some degree a deep-seated abhorrence and indeed fear of nuclear war. Those who participate in the new spirit react in a constructive way, finding hope in the mutual interest in survival that we share with the Soviets. But the spirit has not yet reached a great mass of hard-working people dependent on jobs that may be seen as part of the national economy fed by the arms race. Many of these and others react to their fear by psychological withdrawal from reality, grasping eagerly at the straws of false hopes, particularly if these come from the perceived authority of a president. If the spirit of moderation can be kept alive and growing, reaching ever more of those not yet convinced, perhaps it is not yet too late for replacement or conversion of hawks by doves in Washington to lead to a genuine try that may still succeed at least in drastically reducing the level of arms.

Three decades or so ago, when the first of these papers was written, the idea of living under a nuclear sword of Damocles was so new and alarming that some of us thought it rather unlikely that we would have more than two or three decades without nuclear war. The two decades between world wars I and II seemed to have set the pace. There was a special urgency in the plea for arms limitation. We may now be very thankful that the worst has not happened, that there seem to be no such national ambitions as led to world wars I and II, and that deterrence has at least succeeded in keeping local wars below the nuclear level. During these decades without calamity, complacency has crept in, and that is part of the danger. We must appreciate the fact that three decades of successful bilateral deterrence proves little about the stability of future deterrence, either bilateral or multilateral. So far we have survived in the presence of terrible and destabilizing weapons, but there will be even more of them, and more destabilizing, if we do not stop. If the human race is destined to survive the crisis of this transition into the nuclear age, we have yet to achieve a safer path than perpetually increasing armament.

Throughout these thirty years there have been periods of intense discussion of various special aspects of the arms race, such as shelters for civil defense, counterforce or ''no-cities'' strategies, ABMs, or the test ban. Some of the articles in this and other chapters remind us that, as we hear these ideas proposed anew in very recent years, they are not new and that many of the arguments adduced long ago are as relevant as ever today.

We Haven't Really Tried

[1955]

It is generally recognized by those who give any thought to the thermonuclear armament race that each side will have the capability of dealing a devastating, almost annihilating, blow to the other by about 1957. Military strategists who until quite recently persisted in thinking in terms of continuing our effective atomic technical superiority indefinitely now talk in terms of "stalemate." This is essentially an optimistic term, implying the hope that the "mutual deterrence" of so overwhelming a threat will prevent either side from making demands upon the other. One shrinks from thinking what will happen if one side should continue with bold moves, for the other must then either retire or stand as boldly as it dares short of surely setting off the conflict, and here we would have the pattern that has inaugurated the comparatively minor world wars of the past.

Statesmen have given lip service, at least, to the need for international agreement to take the place of the armament race. There are formal diplomatic developments that lead some to hope that we are making progress. Yet we have been receding farther from any effective arms limitation or disarmament agreement all the time. For the inherent difficulties in devising any effective agreement have been mounting with the size of the stockpiles of destructive potential, and at such a rapid rate that the diplomatic developments would have to be fast indeed to keep pace.

The steps we have taken to meet the demands of the atomic age have been half-measures, or less than that. For the present it may seem better to be taking half-measures than none at all, and we rightly applaud them mildly, but if they fall short of preventing race suicide in the end, it will be small comfort to have taken them. Among those responsible for the military strength of the nation, there has been an alarming tendency to place all our eggs in the striking-force basket, with disproportionately little thought and money for developing the means by which to protect the bases, let alone the rest of the country. There has been some mild encouragement for industry to build new plants in rural areas; yet we continue to build skyscrapers in obsolete cities, and families continue to ignore the problems of flight and shelter. Our government has at times recognized that unity of the free world is essential to our long-term resistance to Soviet expansionism, and that poverty in underdeveloped areas is a cause of world dissension, yet our Marshall Plan and Point Four aid have been localized and short-lived.

In response to the world's yearning for better things to come, President Eisenhower launched his atom pool proposal which is now progressing in its limited sphere. The plan is dramatic because it is both international and atomic, and so appeals to the popular imagination that the Soviets appear ready to go along. The proposal is for the atomic "haves" to contribute crucial materials and at least some limited amount of technical information to the "have-nots," but because there has not yet been proposed a sufficient pooling of experts as well, the first goal will have to be the gaining of reactor experience through construction of small pilot or research reactors, and at this pace the more needy nations will not receive the benefits of atomic power abundance until long after the more advanced countries. The discussions involved in international cooperation on the research atom and the industrial atom may incidentally lead to deeper discussion of other atomic and world problems, but it's a long way from here to the actual control of the military atom. The atom pool plan is thus part, and indeed the most hopeful part, of the diplomatic progress which is so slow that, in comparison with the deteriorating atomic military situation, we are not even holding our own.

In seeking effective disarmament, too, our efforts have been half-measures at best. The Soviet decision in 1946 to oppose the Baruch plan, and the frustrating behavior they have employed to carry out that decision, have been our excuse for not doing more. Of course, we cannot make an agreement with a country that is determined not to agree. But we have generalized from the failure of one early proposal repeated through one period of time, to say that the Soviets will never (without a fundamental change of government) agree to any sensible plan. Faced with a definite refusal of this sort, reiterated with annoying insistence, we might choose between two courses. One is merely to wait for some sign that they have changed their attitude toward the general problem—to "keep the door open for negotiation"—and this is what we have done. The other course would be to continue also to try to induce them to change their minds, with the imaginative perseverance characteristic of Yankee salesmanship. To consider it too much trouble because the effort might be futile, as we apparently have done, is to underestimate the importance of solving the armament problem without war. We could make a much more exhaustive effort to devise a plan that would make them change their minds. It is to their advantage, just as it is to our advantage, to avoid serious risk of atomic war. In principle, it should be possible for us to get together on a mutually advantageous deal to mitigate the risk, and we should be searching for an effective proposition. In practice, there are difficulties, and we have not made a real resolute try to overcome them.

To the concerned citizen who feels overwhelmed by the hush-hush complexity of atomic problems in general, it may seem natural to assume that the government is doing everything possible to find favorable ways out of the armament dilemma. Even though the "keep the door open" policy is all that appears of disarmament endeavor on the surface, he might assume that a deeper search has been made in secret. The release of the Gray Board testimony by the AEC last June, and associated unofficial releases of official information in the past year or two, have revealed by implication how little has been done. Through these windows we glimpse not a calm, rational approach to the problem of trying to find ways of avoiding an eventual jittery "stalemate" in which there can be no confidence of avoiding for long the fatal mistake. Instead we see interdepartmental bickering over the means to be used for immediate military strength, a rough-and-tumble struggle that apparently precludes calm thought about our long-term welfare. In a bitter atmosphere of suspicious competition and ready accusation of traitorous motive, there has been little room for taking seriously any subtle thoughts about procedures that might foster international agreement.

Thus without the kind of thorough and imaginative preparation that might make possible a determined effort at successful negotiation, our delegates to the UN continue to meet the routine needs of atomic diplomacy as it unfolds in a familiar pattern. At a critical time when the West seemed to be making progress in settling the question of West German rearmament, the Soviets have come forth with new concessions concerning disarmament, this time abandoning insistence on prior prohibition of atomic weapons but still carefully leaving a broad gap between the stated positions of the two sides. With the meaning of their conciliatory words still unclear, negotiations in the UN Disarmament Commission are getting started again.

Secretary Dulles has said that if the framers of the UN Charter at San Francisco in the spring of 1945 "had known that the mysterious and immeasurable power of the atom would be available as a means of mass destruction, the provisions of the Charter dealing with disarmament and the regulation of armaments would have been far more emphatic and realistic." The tenth year of the UN, the year 1955, is the occasion for calling for revision of the charter, which became obsolete the year it was written. The technical difficulties of atomic control have become so acute in this single decade that already no revision appears to be meaningful short of establishing an unchallengeable UN military arm, complete with atomic armaments to cover the uncertainties of retiring national stockpiles, and with appropriate executive arrangements to assure against misuse. To propose this, whether or not others may oppose, would seem to be in the interests of a country whose cherished tradition has been the development of government to protect its people from oppression. Yet there are political signs that we are far from ready for any such bold step.

Between the strong extreme of real disarmament on the one hand, with its most desirable effect of mitigating the danger and its great difficulties associated with the necessity for minute inspection, and on the other hand, the weak extreme of the atom pool proposal that does not directly affect military affairs, there lies the possibility of placing the testing of nuclear weapons under international control. This could have a direct and beneficial effect on the problems of the military atom and has the great advantage that it does not require detailed inspection to interfere with internal national affairs. In its mildest form, control of testing could be aimed merely at alleviating the international difficulties arising from the spread of radioactivity, recognizing that the tests are a matter of international concern by placing the gross conditions of testing under international control, perhaps on a balanced quota basis. This need not appreciably hamper weapons development or technical secrecy and could have a benign effect on international relations. A stronger form of test control would include a ban on testing of weapons over a certain power, perhaps just H-bombs, or perhaps H-bombs and large A-bombs, or perhaps all nuclear weapons. This would have as its main purpose the slowing down of development of techniques of offense equally on both sides, giving the defense more chance to be effective and thus reducing the powder-keg explosiveness of the future world: a very real and tangible advantage which should be apparent to both sides.

Many of the measures or half-measures that reduce tensions or foster sympathetic attention to the problem of terminating the arms race are worth pursuing in the hope of improving the world political climate, making a fatal flare-up less likely and eventual far-reaching agreement more likely. But the test ban or test control proposal provides an opportunity for real progress now on the international military atomic threat, and action on this low level, for which world political evolution seems about ready, would be worth much more than many words about a higher level of agreement for which we seem unready.

Shelters and the Chance of War

[1962]

Most discussions of the efficacy of a shelter program confine themselves to answering the question, "If war comes, will a shelter save me?" or, better, "If a certain type of nuclear attack should occur, would a specified type of shelter program reduce the likelihood that an individual would be killed?" While there has unfortunately been much irresponsible exaggeration of the effectiveness of a shelter program in providing "safety," one responsible conclusion frequently reached is that simple fallout shelters would indeed save many lives in the event of an attack directed only against military bases, and would save many fewer lives in more widespread attacks. This is not the only consideration needed to determine whether a fallout shelter program would actually save lives. It is at least equally important to determine whether the existence of a serious civil defense program will increase or decrease the likelihood that war will occur.

The danger that a person will be killed depends not only on the capabilities of shelters but also on whether there will be a nuclear war. The probability that a person will be killed by nuclear war is the product of the probability that such a war will occur, *multiplied by* the probability that a person will be killed *if* war occurs. Both of these factors depend on whether we build shelters, and what kind. Assuming that the probability of being killed *if* nuclear war occurs is decreased by shelters, there are reasons why the probability that nuclear war will occur is increased by an extensive shelter program. Although it is difficult to estimate these influences very accurately, and other factors need to be taken into account, the net result seems to be that a substantial shelter program will increase a person's probability of being killed.

The types of shelter programs we consider might start with no program at all; then go to a simple program of home food stockpiling, perhaps supplemented by stockpiling of cement blocks with which to improvise a lean-to shelter at the last minute, for the sake of hiding from fallout in basements; then backyard shelters; then community fallout shelters and stockpiling; and so on up to an enormous national program of deep underground firestorm shelters, complete with internal air supply and rigid plans for postwar controls. As we come to consider more extreme and costly types of protection, involving essentially living underground in peacetime so as not to be caught outside by surprise, the loss of civilization (including cherished political institutions) becomes the more serious part of the loss, until we consider an attack so severe as to deprive the survivors of all means of supporting life.

The problem now is to decide which is more in the national interest, no shelter program at all, or a limited one aimed mainly at fallout protection from limited war. However, this question cannot be completely isolated from the effects of a very extensive shelter program, because an initial program of simple community fallout shelters will soon become obsolete and will lead to demands for blast and firestorm shelters for city dwellers in order to try to protect all citizens equally.

Influence of CD on the Likelihood of Nuclear War

The reasonableness of a serious civil defense program is closely tied in with the extent to which we can bring the arms race under control and emerge from it without global nuclear war. If there is no hope that we can do anything but continue the unlimited and unending arms race to the bitter and fatal end, then perhaps the sooner we dig in deep, even going so far as to limit population or selecting a special group for survival to make practicable the storage of long-term supplies, the better. Or if the life of a mole is so unpleasant that we cannot tolerate maximizing the chances of national survival in a first nuclear encounter, then the sooner we make more modest preparations which might be helpful in case of a limited attack, without interfering too much with normal life in the meantime, the better. Even though such a course might, by seeming provocative, bring on the unwanted nuclear war sooner, this could not be considered such a great calamity if nuclear war were inevitable in any case.

But the situation is not that bad. There is hope that we may think and work our way out of the perpetual uncontrolled arms race, and the extent of the hope is influenced by our decisions about civil defense.

If there continues to be a proliferation of refinement of nuclear weapons and of the number of nations able to threaten their use, nuclear war will be inevitable. The only long-range hope to avoid eventual nuclear war is through some form of arms control. This in itself is not sufficient—other world problems will have to be solved eventually—but it is a necessary part and will help provide time for the rest. The nuclear danger is not simply going to fade away. The effect of a serious shelter program on work for a stable peace (and thus on the likelihood of nuclear war) may be presented in four main categories: the direct influence on our negotiating position, and influence through public opinion, diversion of effort, and vested interests.

Influence on Negotiating Position

There are many, including this writer, who believe that our sights must be set on far-reaching disarmament, to be attained, for instance, by way of the region-by-region disarmament mechanism and an interim low-level transitional deterrent stage. Such procedure postulates successful negotiation and formal agreement, of which there has been lamentably little in the past, perhaps because the goals and possible rewards of past negotiations have been far too limited.

There are others who favor instead immediate attention to much more limited arms control, perhaps to be attained by a sequence of approximately reciprocated unilateral actions. The immediate aim of such a program would be to reduce the likelihood of accidental war by placing major emphasis on relatively invulnerable weapons systems not requiring instant response to suspected attack.

A still more modest proposal is a nuclear weapons test ban agreement, aimed not at actual arms control, but at control of the development of future arms and at preventing their spread to other countries. It is the only type of significant arms limitation proposal which the United States has submitted in a reasonably complete and carefully considered form, and even here our negotiating position has been faltering and only intermittently enthusiastic and constructive.

Many people are too glib to say that there's no use thinking of arms control because the USSR won't agree to anything that is not very much against our interests—as though our timid and preliminary past negotiations on the edges of the disarmament problem had thoroughly explored the possibilities of arms limitation.

In the quest for any limitation of armaments, the decision to adopt a serious civil defense program could be and probably should be used as an important bargaining item.

The real problem is of course to find a program which is in the interest of both sides because it genuinely and appreciably reduces the likelihood of nuclear war without significantly changing the cold war balance. The more advantages that can be attached to such a program, the more likely it is that it will seem acceptable to each side. If any kind of an agreement can include freedom from the necessity of disrupting our way of life with a serious civil defense program, because the other side is refraining from such a program, this feature will make the agreement more attractive. It will increase the chances that the USSR will agree to an arms limitation scheme having great value to us as well as to them.

In the course of negotiations for a formal arms limitation agreement, the no-shelter proposal could be both a threat and a promise. It might be made apparent that we are refraining from a shelter program out of hope for agreement and that if the negotiations fail, we will feel obliged to take the next serious step in shelter construction.

Mutual abstention from large-scale civil defense preparations would also fit in well with a policy of reciprocated unilateral acts. Quite by itself, it could constitute a useful first step in practicing the art of reciprocal restraint in preparations for war. That is, we might reasonably propose to the USSR: "We are not starting a large-scale shelter program and won't as long as you don't." Compliance would be easy to observe (the USSR is no longer almost completely closed to foreigners), and the scheme would have a nice balance to it—neither side would be trying by this means to evade the effectiveness of the other's deterrent. . . .

Public Opinion

The extent to which public opinion will tolerate and encourage a forward-looking foreign policy, departing from the prenuclear acceptance of actual war as a normal extension of diplomacy, depends among other considerations on an awareness of the very real and imminent danger of nuclear war and on people having some doubt whether the threat of nuclear war will be an effective means of avoiding it. The initiation and public acceptance of a shelter program may influence public opinion about foreign policy in various ways.

Public reaction to the stepped-up discussion of shelters during the past year demonstrates that even just thinking about concrete activity close to home increases awareness of the problems of the nuclear age. If such a program were promoted exclusively by completely frank and honest advertising, with due recognition of the uncertainties of the future nuclear dangers and the haphazard nature of the protection afforded by even the best of shelters, it could be an influence toward a more flexible foreign policy. If we are in fact going ahead with a public shelter program, it is most important that it increase public awareness of the need for an alternative to perpetual war preparation.

During the first months of the stepped-up program, the advertising was preponderantly quite misleading, and of a sort to develop a risky false sense of security through overconfidence in shelters, particularly family shelters. The new official booklet *Fallout Protection* is at least a big improvement in this respect. While it still goes too far toward allaying fears by claiming that our foreign and defense policies "make such an attack highly unlikely," it is frank in the way it indicates that "the experience would be terrible beyond imagination and description" and realistic in many of the elementary suggestions it makes about behavior in that dreaded event. (However, its deemphasis of the firestorm danger seems unfounded.) When we consider financing a big public shelter program, it seems quite unlikely that it will be possible, in our usual way of doing things, to "sell" the shelters to the taxpayer without greatly exaggerating their effectiveness and minimizing the problems of postattack recovery. The likelihood of the consequent overconfidence in shelters and underemphasis of the horror of nuclear war is one of the disadvantages of such a program.

The founding fathers wisely put the decisions of diplomacy, even including the initiation of war, in civilian hands. Civilian thought has since been molded to a considerable extent to conform to military thought by the fact that large numbers of civilians have had military experience. An extensive civil defense effort will probably bring many more people into a type of special training very similar to military training and indoctrination. This will tend to increase military dominance of civilian thought and decrease the freedom of action of our policymakers in seeking nonmilitary solutions of the nuclear age dilemma. The psychology of the garrison state does not promote spontaneity in the constructive and good neighborly aspects of foreign policy.

When a man starts shopping for a new car, he may be very objective and may compare critically the virtues and faults of various makes. Once he has made the close decision and bought one, he becomes a strong partisan of that make. Psychologists consider that he identifies his ego with his decision, and tends to exaggerate the virtues of his car to justify his choice.

Similarly, if a man participates in a community decision to build a community shelter, and particularly if he builds a private shelter in his own basement or backyard, he will tend to identify himself with his decision, and will probably, perhaps unconsciously, favor a foreign policy which makes shelters look necessary. He would have reason to fear looking foolish if "peace should break out." This reaction would take place on so large a scale that national legislators and other policymakers could not help but be responsive to it. A closely allied reaction which does not require the same degree of personal involvement is a feeling of resignation to the inevitability of nuclear war in the face of extensive preparations for it all around.

Diversion of Effort

The more strongly nations become convinced of the nuclear age truth that all-out war is obsolete as an instrument of national policy, the more will the world competition for influence become economic and cultural rather than military. What we need is a national dynamism toward constructive ends.

A shelter program will detract from all these goals and a massive one might detract disastrously. It will compete for money and materials; it will compete for initiative and talent; it will compete for the devotion of men. Some funds which should be applied to increasing teachers' salaries and replacing substandard schools, for example, will go into school shelter construction. With new demands on the public purse, and a renewed spirit of self-sufficiency engendered by digging in, it will be more difficult to get appropriations and enthusiastic personnel for aiding economically backward nations.

As already mentioned, our decision to initiate a large shelter program will probably induce the USSR to do the same. Since they will thus have a similar diversion of effort from their side of the economic and cultural competition for world influence, the fact that we are diverted from it may not influence the outcome of the competition as much as its nature. It seems that the world would be far safer from nuclear war if the trend were toward emphasis on the economic and cultural aspects of the competition rather than toward a perpetual confrontation of two great garrison states, with others developing as nuclear weapons spread to other countries.

Vested Interests

If anyone doubts that vested interests play a large part in determining the nature and magnitude of our national spending for armaments, he should hark to Representative Whitten, a Democratic member of the House Appropriations Defense Subcommittee.

> I am convinced that defense is only one of the factors that enter into our determination for defense spending. Others are pump-priming, spreading the immediate benefits of defense, taking care of all the services, giving all defense contractors a fair share, spreading the military bases to include all sections.

Practically every senator and congressman is under pressure from some constituents whose livelihood or prosperity is supported directly or indirectly by defense expenditures in his district. This pressure primarily seeks sustained and increased local expenditures, but also tends to be opposed to the parts of foreign policy aimed at reducing tensions.

The adoption of an extensive civil defense program would bring a whole new class of workers and industrialists into this picture. Large segments of the building trades and construction industry would be expected to join in the chorus favoring a tense and short-range interpretation of national interests in general, and an overemphasis of brinkmanship in particular. Even the preliminary talk about home shelters has produced a flurry of shelter construction contractors with an economic stake in the national decision.

Influence of Shelters on the Nature of an Attack

Current doctrine, on which a shelter decision may be based, is that an attack, if it comes, will probably be directly primarily or exclusively against Strategic Air Command and missile bases in an attempt to minimize retaliation. This presumes that an adversary mad enough to attack will be sane enough to attack rationally. Then too, an adversary's motives and rationale cannot be predicted with confidence. It seems likely that, if an attack should come, not by escalation from an accident but in cold blood, it would be planned to hurt us so badly as to make us give in to some demand, while doing what it could to minimize retaliation. Even if there are not enough missiles to hit all bases, it seems likely that at least a few missiles will be diverted against cities to make the attack hurt enough to try to induce submission to demands. If we have shelters, a few more missiles would be required against cities than if we do not. Thus, the net effect of shelters may be to divert some missiles from counterforce to counterpopulation, that is, to spare some of our missiles rather than our people and thus to increase our retaliatory capacity slightly. The effect of the adversary's shelter program would be to make our blows less effective and thus to reduce our retaliatory capacity. Here too, the shelter programs tend to compensate one another, and both sides would have built shelters without saving any lives.

In the farther future, when numbers of available missiles will be much greater, the question will be how massive an attack will be required to make a nation like ours give in. An enemy's attack requirement will depend on how effective a shelter program we have. If the shelter program is very effective indeed, the enemy may be forced to make an attack of the sort to deny us any means of livelihood after emerging from the shelters or after exhausting the supplies in them. It will be possible to upset the ecology completely, to burn all forests (promoting alternating floods and droughts in all rivers) and to poison all agricultural lands with radioactivity. Thus the long-term end result of an arms race coupled with a shelter race will be that nations will aim at complete destruction of each other in nuclear war; while leaving populations sufficiently exposed can permit a war to stop with some life still possible thereafter.

Consideration of this extreme case suggests both that the effectiveness of shelters will be circumvented by modification of the attack in almost any situation and that, rather than to make a vain attempt to escape by digging in, we had better be devoting our energies to building a better world and to finding political ways to exploit the common interest of all men in continued life.

Conservative Judgments
and Missile Madness

[1968]

The crux of the present stage of evolution appears to be this: that the great nations are judging their military security requirements so conservatively that the arms race is in the process of taking civilization over the brink of the nuclear abyss. If nations would act on reasonable judgments, and not on paranoically conservative judgments, it should be possible to stop and even to reverse the nuclear arms race without reducing, and in fact increasing, the military security of each of the nations involved. Tragically, mankind is not following the reasonable course.

The unprecedented threat of nuclear weapons cannot be met with judgments based on traditional military thinking. In the past, getting there "fustest with the mostest" was the conservative and reasonable military tactic. With strategic nuclear weapons the great nations cannot hope to win a major war, but can hope through deterrence to prevent it. The potential destruction is large, compared to what might be considered "acceptable," by so wide a margin that it is not necessary to have more weapons than an adversary has, or even as many, to deter an attack. "Fustest with the mostest" is no longer a reasonable requirement in the nuclear age, but by tradition it is still the conservative requirement. As long as two nations are practicing this kind of conservatism, each attempting to sustain nuclear superiority over the other, there can be no stopping the race.

As recently as a year ago the publicly available information made it appear that the arms race was tapering off. Our installations of land-based and sea-based strategic missiles were essentially completed and we seemed content with their numbers. The Soviets were installing new strategic missiles at a relatively low rate and appeared to be satisfied with a missile force perhaps one-third or one-fourth as numerous as ours. It was conceded that they could retaliate terrifically if we should be so foolish as to attack them first, and this fact still provides an example of the stability of the deterrent; superiority or near-equality is not needed.

But suddenly, with the announcements of ABMs (antiballistic missiles) and MIRVs (missiles with several independent warheads), the arms race seems to be starting to spiral madly upward again. The most dramatic announcement was made by then Secretary of Defense McNamara on September 18 in San Francisco. He discussed ABMs at length, but MIRVs only by implication, as one of the means by which we had already arranged to penetrate present and future Soviet ABM deployments. He reluctantly announced the administration's decision to deploy a "thin" antiballistic missile defense. He expressed his hope that this would not be the beginning of another dangerous and debilitating round in the arms race, but said he feared it would be.

McNamara's performance in Washington as secretary of defense was remarkable for the integrity with which he sought to stretch the taxpayers' dollar by reducing duplication in the armed forces. This made him unpopular with many of the military men over whom he presided and with many of the defense industry–connected congressmen whose enormous military appropriations he did or did not spend, largely as he saw fit. Indeed, it has been suggested that his unpopularity with the military chiefs, even more than his restraining influence in Vietnam or his disappointment over losing his battle to save us from the ABM burden, was responsible for his resignation from the cabinet.

Nuclear Superiority or Near-Equality?

Secretary McNamara's now famous September 18 speech was remarkable as much for its analytic tone as for its substantive announcement; for the insistent and yet conservative way in which he explained that we do not need a Soviet-oriented antiballistic missile system while he announced that we are nonetheless to have a "Chinese-oriented" one. His discussion of first-strike capability and second-strike capability was well defined: First-strike capability implies that one nation has a strategic nuclear striking force so overwhelming as to be able to destroy an adversary's retaliatory forces and still inflict tremendous industrial damage—damage that would be "unacceptable" in any meaningful political sense. With such first-strike capability, a nation could attack without fear of immediate military reprisal. Second-strike capability, the basis of deterrence, implies the ability to strike back and inflict unacceptable damage even after an enemy has tried to wipe out the retaliatory forces. With the authority of inside knowledge, McNamara assured us that neither the United States nor the Soviet Union has or can hope to develop a first-strike capability against the other, but that both have and with reasonable vigilance will preserve a second-strike capability.

More specifically, he went on to say: "By using the realistic measurement of the number of warheads available, capable of being reliably delivered . . . I can tell you that the United States currently possesses a superiority over the Soviet Union of at least three or four to one. Furthermore, we will maintain a superiority. . . .'' But he did not say that we will, or need to, maintain as great a superiority as we now have. On the contrary, he said that our superiority "is both greater than we had originally planned and is in fact more than we require" and is of limited significance because "the Soviet Union could still—with its present forces—effectively destroy the United States, even after absorbing the full weight of an American first strike.''

Here again we find it recognized from within official circles that not even near-equality among adversaries is necessary to make deterrence effective, so clearly there is no need for superiority.

More recently, information has been released that the Soviets have accelerated their ICBM deployment within the last year or two and concern has been expressed on the campaign trail and elsewhere that they are about to overtake us. In a report to Senate committees on February 1, McNamara disclosed that the Soviets' deployment of land-based ICBMs had increased from 340 in 1966 to 720 as of October 1967. He estimated that this, with allowance for sea-based missiles and for bombers, still leaves the United States ahead in numbers of deliverable warheads by a score of 4,500 to 1,000, a ratio even somewhat greater than he had quoted on September 18. The U.S. score is apparently swelled by the multiple warheads of the new submarine-based Poseidon missiles and the trend toward MIRVs will probably continue to swell the numbers and complicate the future balance. The Soviets have probably long known of the prospect of our Poseidon multiple warheads and the upturn in their ICBM deployment may be their response. Thus the virtual tapering off of the arms race apparent over a year ago seems to have ended, but the significant fact remains that the Soviets, in reacting only enough to keep the score around four to one, seem to be indicating that they are willing to settle for a substantial degree of strategic nuclear inferiority as their basis for deterrence.

The MIRV development is probably closely tied to the prospect of ABMs: ABMs beget MIRVs and MIRVs beget ABMs. That is, our MIRV program was initiated to penetrate the Soviets' premature small ABM deployment around Moscow and its possible successors; our ABM deployment is probably motivated mainly in response to the anticipation of Soviet MIRVs, following ours. If so, it is an unnecessary response—and the wrong response—for there are other, less destabilizing, options available. We could put greater emphasis on land-mobile and underwater missile deployment, or could forgo the economy inherent in MIRVs by deploying separately enough missiles to carry the warheads singly, so as to present a greater profusion of silos as targets, directing money to the aerospace industry this way rather than via ABMs. The advantages of this option might eventually put an end to the MIRV menace.

McNamara doubtless had these responses and counterresponses in mind as background for his September 18 speech, but he did not discuss them in detail. Nor did he go into the question of the more urbanized population and greater urban vulnerability of the United States. He did recognize the greater average power of Soviet missiles as considerably less important than our greater numbers. These details modify the argument somewhat, but with due allowance for them his remarks still indicate that if the Soviets had the sort of superiority over us that we have long had over them, deterrence would still be as effective as ever. That is one important inference to be drawn from his facts and ideas.

Conservative Judgments

Another important lesson concerns the nature of a conservative judgment in the American establishment. McNamara pointed out that ''the strategic planner must be 'conservative' in his calculations; that is, he must prepare for the worst plausible case and not be content to hope and prepare merely for the most probable.'' He assured his critics that our estimates of Soviet strategic capabilities have been and will remain conservative. In 1961 to hedge ''against what was then only a theoretically possible Soviet buildup, we took decisions which have resulted in our current superiority.'' It is, of course, considered natural that the Soviets, too, should judge their requirements conservatively. The Soviets' conservative judgment apparently permits them the economy of having less than one-third as many deliverable warheads as we have, provided they can give the Soviet people, with their memories of Hitler's invasion, the psychological lift of something that may properly be called defensive in the form of a rather ineffective antiballistic missile deployment.

The Washington view, as reflected in McNamara's speech, seems then to be that conservative decisions to maintain strategic superiority in the ''worst plausible case'' should be the norm. If followed by both sides, this view would permit no end to the arms race. With both nations allowing for the ''worst plausible case,'' they are even more strongly locked into the race. Logically, the feasible way out would seem to be for both sides by agreement to seek approximate equality. Since there are uncertainties in judging equality, even if no information is hidden, some toleration of slight inequality would be needed. However, as McNamara made clear, all that is required is assured second-strike capability on both sides, and for this there is great latitude. A very rough approximation to equality should be enough, perhaps within a factor of two or three either way. This should eliminate concern for the worst plausible case, especially since uncertainties are less now than in 1961. If both sides would take this reasonable attitude, arms control would be easy.

We should consider it fortunate that the Soviets seem content to accept substantial numerical inferiority, provided they have a bit of ABM as a kicker. If we really wanted arms control, this should make it even easier for us. We should be able to insist on a modest degree of superiority—unnecessary superiority as a concession to our traditional conservative judgments— and still stop the arms race in its tracks. In this sense a modest Soviet ABM deployment for internal propaganda might be good. It might permit each side to develop at least a politically useful illusion of superiority.

Until McNamara's September 18 speech, it appeared possible that the nations would follow some such sensible course. Now it appears much less likely; after holding out for many months, he had to give in to pressures to start the deployment of at least a "thin" ABM system. He insisted that it is merely a China-oriented system and eloquently pleaded that it should not become more. But are these entreaties more apt to be heeded than his last, particularly since his departure and in the face of all the political pressures? His congressional and military critics and industry—the "military-industrial complex"—have been given an inch. It seems too likely that they will take a mile. For, as Mr. McNamara so aptly warns, "There is a kind of mad momentum intrinsic to the development of all new nuclear weaponry" and "strong pressure from many directions to produce and deploy the weapon out of all proportion to the prudent level required." This seems to be the kind of warning we can hear from a man experienced in high office when he is on the verge of leaving it. It was in his farewell address that President Eisenhower warned against the "undue influence of the military-industrial complex."

It is impressive how quickly, after the September 18 announcement of a "thin" system, the pressures built up in Washington for the deployment for a massive, Soviet-oriented ABM system. The very next day, Senator Henry M. Jackson (D., Wash.) was hailing the decision as a step toward a massive ABM system and in November he had his Subcommittee on Military Applications (of the Joint Committee on Atomic Energy) hold hearings trying to discredit McNamara's views. As if to whip up public and congressional enthusiasm for an ABM race, technical information was soon released about

X rays and neutrons as missile destroyers and about the problems of U.S. and Soviet missiles carrying X-ray shields. These are special aspects of the eternal countermeasure seesaw that are inevitable if we don't stop somewhere. It is too easily forgotten in the fragmentary public discussion that McNamara had access to all such matters in making his judgment of the ineffectiveness of an antiballistic missile system against massive attack.

In stating that "we have decided to go forward with this Chinese-oriented ABM deployment" he was really announcing an administration decision to go ahead and spend limited funds that had already been appropriated by Congress. The decision seems to imply that the president considered yielding to pressures for at least a "thin" system to be good politics for an election year. However, most of the $5 billion estimated cost of the "thin" system has yet to be appropriated, so that Congress could still have final say and stop the deployment, if political miracles were possible.

Actually, the initial funds are available because Congress tried to push the administration into this. In August 1966, Congress passed a supplemental appropriation, beyond the budget requested and until recently not spent by the administration, of $154 million to start long-lead-time items for ABM deployment, and only fourteen senators voted against it. There was also another standby appropriation about twice as large, but there may be more opposition now. Discussion of international implications that has since highlighted the question has probably impressed more senators by now. Others will find it difficult to fit the large present and projected costs (does the estimated $40 billion estimate for a massive system really mean $100 billion?) into a war-swollen budget. There is resentment of the use of the Tonkin Gulf Resolution as a blank check and fear that a "thin" system may amount to a blank check for a massive ABM system and for an unpopular shelter system beyond that. The limited test ban treaty followed the Cuban missile crisis. The political miracle of a congressional "about-face" is not impossible.

The Chinese Excuse

The Chinese orientation of the thin ABM system is probably largely an excuse not widely believed by Washington officialdom. As Senator Clark of Pennsylvania said in the Senate on October 9: "I think it is perfectly clear that practically everybody except the military-industrial complex, which would profit from the building of this system, is of the view that to build the system against the Chinese, realizing it is not good against Russia, just does not make any sense at all." McNamara was at his least persuasive in announcing our

China-oriented system, as though apologizing for doing so. After saying "there is ample evidence that China well appreciates the destructive power of nuclear weapons" and "has been cautious to avoid any action that might end in a nuclear clash with the United States," he pointed out that "it would be insane and suicidal" for her to attempt a nuclear attack on the United States or our allies, but she might miscalculate. Then came his announcement: "And since, as I have noted, our strategic planning must always be conservative, and take into consideration even the possible irrational behavior of potential adversaries, there are marginal grounds for concluding that a light deployment of U.S. ABMs against this possibility is prudent."

Actually he had not noted anything about "possible irrational behavior." On the contrary, his previous dissertation had dwelt exclusively with the need for a deterrent against rational attack, in the form of a second-strike capability against the Soviet Union. We trust the Soviet Union to be reasonable. In this we have no choice, a first-strike capability being unattainable. We are comforted that Russia has evolved from the days of Stalin. But now, to replace Stalin as the Great Unknown, the Ogre, we have preserved China, carefully nurtured by years of exclusion as a political untouchable. Even though we will for a long time have not only a second-strike but even a first-strike capability against her, we claim to need further insurance against her irrationality. This is a ridiculously disproportionate assessment of the dangers.

Avoiding Proliferation

Among the sources of serious trouble ahead are nuclear proliferation and widespread starvation from overpopulation, on the time scale of a decade or two, and depletion of natural resources on a longer scale. In facing these we seem to be making recklessly careless judgments of diplomatic needs while making ultraconservative judgments of military needs. A conservative judgment of the need to meet any one of these troubles would come directly into conflict with the so-called conservative judgments of military needs that are taking precedence.

Of most immediate diplomatic concern is the nuclear proliferation problem. Among the arms controllers of the State Department and the Arms Control and Disarmament Agency (ACDA) it has a high priority, for our government is fully aware of the magnified uncertainties there will be in trying to make deterrence work dependably in a world with many nuclear nations. Having China and France as nuclear nations is bad but now inevitable, because we did not give the proliferation problem high enough priority soon enough. Having many more nuclear nations will be much worse and one branch of our government is trying without much support to do something about it, something that will also have a chance in the long run of mitigating the Chinese-French part of the difficulty by reducing the incentives for these nations to spend heavily on nuclear arms.

A nuclear nonproliferation treaty has been the goal of the Eighteen-Nation Disarmament Commission meeting at Geneva for several years. At it, our negotiators have been patiently trying to reconcile the limitations of conservatively judged U.S. and NATO military needs with similar limitations placed on the Soviet negotiators. There has been progress, but it has been lamentably slow. The real trouble now appears to be that the lesser powers are reluctant to sacrifice future nuclear options when the great nuclear powers are required by the treaty to give up very little—essentially just the freedom to supply nuclear weapons to nonnuclear nations.

The failure of the major nuclear powers to make more concessions in limiting the arms race could easily be blamed on Soviet reluctance to admit inspectors; that overemphasis on secrecy is one aspect of the Soviets' conservative judgment of military needs. But the failure to make concessions can be attributed just as directly to U.S. conservative judgment of military needs. By judging more realistically the relative importance of our diplomatic and military needs, we could initiate concessions that could break the deadlock.

Unfortunately, our recent decision to mount a ''thin'' ABM defense is a move in the opposite direction. It indicates how hard it is for us to approach a spirit of military restraint for the sake of broader goals. The offer, announced by President Johnson on December 2, to open our nonmilitary nuclear installations to the same inspection applied to the nonnuclear-weapon nations in a nonproliferation treaty, is a modest step in the right direction, but, not being a matter of military restraint, is apt to prove insufficient.

A reckless judgment of diplomatic needs in the proliferation problem might include the optimistic guess that, if ten years from now nuclear war should break out between Israel and Egypt, for example, the United States and the USSR would not become involved so it would be no real danger to us. If, on the other hand, we were to judge our diplomatic needs nearly as conservatively as we do our military needs, with the worst plausible case in view, it would be a matter of essential policy not to let the world evolve toward the possibility of nuclear war breaking out almost anywhere and spreading.

If we had been guided by conservative judgments of diplomatic needs, we would before now have arrived at agreement with the USSR on the terms of a complete nuclear test ban and be in a position to include this in the nonproliferation treaty as a very real measure of restraint by the major nuclear powers.

Destabilizing the Arms Race

The Chinese orientation of the announced decision to mount a "thin ABM," insofar as it has any validity, is a case of trying to judge by the "worst plausible case," namely, the case that the Peking government should be completely insensitive to our deterrent striking power and might either destroy an American city or blackmail our foreign policy by threatening to do so. The thin-ABM advocate knowingly asks: "How would you like to be an American president and have to decide what to do, without ABMs, in the event of a Chinese threat of strategic missile attack on an American city unless we get out of Taiwan?" But this is merely the "worst plausible case" against which we know how to spend a lot of money (always assuming, of course, that mastery of Taiwan is vital to our security). A still worse case would be such a threat accompanied by a claim or evidence that the Chinese have several H-bombs assembled in American cities from "suitcase delivery" of parts. With this as a still worse plausible case, the thin ABM will have accomplished very little for President Johnson in his hypothetical crisis. The deterrent power of our strategic missiles will give him much more comfort.

But the massive ABM race with the USSR, which is likely to follow, will make it almost impossible to provide adequately for the "worst plausible case." Each side will know or guess that its own ABM system should intercept, say, 10 percent—or perhaps even 30 percent—of the incoming missiles in a massive attack. But neither side will know that the ABMs of the other side may not be much more capable. A conservative judgment of the "worst plausible case," if our 1961 estimate of Soviet missile power is any guide, might allow that the other side could intercept 90 percent of incoming missiles. Thus, because of the great uncertainty of effectiveness of ABMs, the conservative judgment of each side might call for many times the missile power of the other. If we should conservatively judge that the Soviets' ABMs would let through only 10 percent of attack missiles and our ABMs 90 percent, then according to this "worst plausible case" we would conservatively require nine times as much attack power as theirs to be about equal in effectiveness. These figures must be an exaggeration, but the example illustrates the destabilizing influence of antiballistic missiles on the arms race.

If it should lead no further, the "thin" ABM deployment is ill advised because it is essentially useless, wasteful of resources, discouraging to those we would have sign a nonproliferation treaty, and a mild goad to Soviet and Chinese nuclear procurement. In itself it may not be fatal to future policy. The greater danger is that it will lead directly to a massive ABM deployment and subsequent missile madness, but the die has not yet been cast. If the growing political and industrial pressures can be resisted, it may still be possible to follow a course of reason.

Conservative Diplomatic Judgment Needed

Toward the end of his speech, McNamara said: "Let me emphasize—and I cannot do so too strongly—that our decision to go ahead with a limited ABM deployment in no way indicates that we feel an agreement with the Soviet Union on the limitation of strategic nuclear offensive and defensive forces is any the less urgent or desirable." Nor is the nonproliferation treaty any the less desirable. A conservative judgment of these diplomatic needs, in terms again of the worst plausible case if we fail to get them, would place them at the top of our national priorities, far above the conservatively judged need to have substantially more strategic missiles than does the Soviet Union. If we were willing to make a trade-off among these conservatively judged needs, we could almost surely achieve the diplomatic goals without sacrificing realistic military needs.

McNamara here speaks of "agreement with the Soviet Union on the limitation of . . . forces." We have, of course, tried diplomatically to reach such an agreement, but have not come to terms. After having produced a surplus of nuclear weapon materials, we have proposed a cutoff on production if they will submit to inspection of their production plants. They object that we want inspection without disarmament and we claim they want disarmament without inspection, and that is essentially where the matter rests. Ours has been a low-priority, no-sacrifice gambit to penetrate the Soviets' dislike of inspection without disarmament. It has been the result of reckless diplomatic judgment, based not on the worst plausible case, but on the best possible hope that multilateral deterrence will work forever and thus that the arms race is fine and there is no pressing need to turn it off.

Thus both reckless judgment of diplomatic needs and ultraconservative judgment of military needs nourish the arms race and are promoted by industry through political pressure because they seem, in the short term, to stimulate industry. If industry could take a longer view, not necessarily so far ahead as the Armageddon at the end of the arms race, but only so far as the industrial prosperity that would follow conversion to a peace-oriented economy meeting the real needs of the world, the pressure might be relaxed and conservative diplomatic judgments might become politically tenable.

Using the ultraconservative yardstick on military needs and the reckless yardstick on diplomatic needs is like double-bolting the front door and leaving the back door open. If we want to distribute the bolts on both doors, if we want to do something reasonably conservative to limit the spread of nuclear arms and change the direction of the arms race, we should start making some diplomatic offers that reflect a more realistic judgment of military needs. If we have three or four times as many deliverable nuclear warheads as the Soviets have, we should, for example, propose to cut off nuclear production on both sides, appropriately inspected, and to make a substantial step toward disarmament by having the United States turn in for destruction four times as many nuclear warheads and their delivery vehicles as the Russians do. If that is not enough, we could afford to offer more, say 40 percent to their 20 percent reduction. Both sides would still have ample second-strike capability and amply deny first-strike capability to the other. Beyond this there are various plans for disarmament by stages widely discussed in the past that should be resurrected for more serious consideration, particularly now that the proliferation problem has become so urgent.

Being numerically superior to the Soviets in nuclear weaponry, we have the option to take the initiative in ending the arms race. Although we blame on them most of the ills of the world, we more than they are in a position to make a generous offer without jeopardy to our deterrent posture.

Our political system has been reasonably adequate in presiding over the domestic aspects of an age of rampant industrial growth for which it was not really designed in 1789. If some special interests have gained unfairly through organized pressure, their profits have found their way back into the common economic stream and a large segment of the population has prospered. But when it comes to making long-range plans for national and world security in the nuclear age, for which the system definitely was not designed, it seems to be woefully inadequate. The world's nuclear problems are too subtle for the average unconcerned citizen and the part most visible to him is the economic manna descending from the defense-industry heaven. This leaves his representatives in Washington responsive to industrial and military pressure groups, but in nuclear-age planning their influence can mean not just short-term special advantage, but long-term population obliteration. Then, too, the interlocking functions of the many parts of a vast bureaucracy provide an automatic veto over almost any distinct departure from established policy.

The groups of concerned, not-so-average citizens seeking a more far-sighted foreign policy lack the resources to compete with this powerfully built-in perpetuation of established trends. They have been trying to forestall the new ABM dimension of the arms race, but the recent "thin" ABM decision shows how little headway they are making. They may act as a brake. They may have prevented its being already a massive ABM system and they must keep trying. But they seem powerless to convince the government of the need for drastic new measures of restraint to attain even a goal so immediate and vital as avoiding nuclear proliferation.

Nuclear Arms Control Aspirations and Frustrations

[1974]

Picture a group of forty men of science in two school buses traveling one fine summer morning in 1945 from Alamagordo, New Mexico, back to Los Alamos where they and their colleagues had invented, designed, and built the first atomic bomb. They had just seen an awesome sight: the first atomic fireball and mushroom cloud. There was a strange mixture of jubilation and sobriety among them that fateful morning. Their three years of effort had succeeded. There was the jubilation of success. But the terrible success showed that nature held no obstacles in the way of mankind's destroying itself. It was now up to man to place political obstacles in the way of his self-destruction. Could he do it, with his propensity to make war after war? It would require developments in international relations as revolutionary and perhaps as rapid as the revolution in the technology of destruction, the culmination of which we had just witnessed in that awesome flash in the desert.

First impressions are very strong. The men who knew that early bomb, and had a gut feeling for what it and its successors could do, took the problem very seriously. So did many others who learned about it from them. They dared think about disarmament, about changing national attitudes enough to achieve a world in which there would be no nuclear weapons. But people soon become inured to evils over which they come to feel they have no control. Perhaps that is how civilizations end. First impressions of the horror of the nuclear threat have faded and pressures for business as usual have prevailed. Not, however, without a long series of attempts to find a new and safer course. Nor without some success in institutionalizing restraint, though ever so slight, in comparison with the need.

The story of these attempts starts with the Acheson-Lilienthal plan that arose out of the concerned thinking of scientists of the bomb project, as developed in committee largely by physicist Robert Oppenheimer, who had led that project. Recognizing that American-British monopoly of the bomb could not last, it proposed giving up that monopoly to an international atomic authority that would keep itself on the forefront of atomic developments and have both the authority and the capability of preventing any nation from making atom bombs. In the interim just after the war and before there were vested interests in nuclear weaponry, this intrinsically wise and generous plan became U.S. policy, but not officially until—renamed the Baruch plan—it had been presented conservatively in the United Nations by elder statesman Bernard Baruch, who appended conditions that seemed calculated to make it unpalatable to the Russians. Under Stalin, they probably would not have accepted the plan anyway. Their counterproposal was to "ban the bomb" as a part of "General and Complete Disarmament," GCD, in which essentially each nation would disarm itself while trusting others to do likewise, a requirement of trust in a suspicious world. These two opposing proposals stood as national policies in the disarmament debates in the UN, essentially as window dressing long after it was clear that each was unacceptable to the other side, and it was even doubtful if it was acceptable to the side proposing it.

While the long-range desire to avoid a very dangerous nuclear arms race dictated that we should, through appropriate international agreement, get rid of atom bombs, there was the short-term reason of economy for keeping our postwar monopoly of them; and military men erroneously predicted that the monopoly would last a long time. The balance of power in prewar Europe had depended on the German army to counterbalance the large manpower of Russia. After the war this was gone and the Russians had many more tanks than the rest of Europe. Immediately after the war, the American army in Europe was the strongest the world had ever known, but it would have been expensive to keep it there in force enough to maintain the presumed balance. The need for the defense of Europe was seen in the light of the Soviets' having made satellites of the countries it occupied at the end of World War II. The atom bomb permitted the economy of demobilizing American military manpower. It was from the very first a deterrent, to deter the Red Army from adventure. The Russians, of course, saw it as a threat against which they had to build a deterrent in kind. This was the background of maneuverings in disarmament circles after the monopoly was lost in 1949, and even after both sides achieved the much more powerful hydrogen bomb in 1953.

By that time the nuclear arms race was on, with preponderant numbers of weapons on the American side. In the early fifties, the numbers of nuclear weapons were presumably barely in the hundreds, but by the end of the decade they were already well up in the thousands. The destructive power of these weapons is enormous. Their most powerful predecessor based on chemical explosive—the World War II "blockbuster"—was a formidable destroyer in itself, twenty tons of TNT. The first A-bombs were a thousand times more powerful, and H-bombs are again a thousand times more powerful still. The great tragedy and human misery of Hiroshima were but a tiny foretaste of what nuclear war could be like.

While the initial brief opportunity to stop the making of nuclear weapons when they were still few was lost, and while they are being found temporarily useful to maintain a balance of power, it came to be more and more generally recognized that they should never be used. Their use would mean that both sides would lose very terribly in war. Thus the main purpose of the opposing and madly growing nuclear stockpiles came to be to prevent their ever being used, although military doctrine on both sides has developed calling for their use at various levels, beginning with battlefield or "tactical" use, in the event of war.

Advances in technology and in numbers on each side have spurred development on the other side. In particular, Sputnik, as a surprising demonstration of Russian technical capability, was followed by a tremendous buildup of nuclear weapons in the United States, giving us a lead of between four to one and ten to one over the USSR at various times through the sixties and until the Russians, reacting to this, began to catch up in most categories in the early seventies.

In a sense, this military buildup has been the main show, and arms control and disarmament efforts have been a bit like a panting dog yapping at the heels of a fast horse. The arms buildup has involved immense technical and industrial activity as well as military organization, and consequently a large vested interest in having it continue. Arms control personnel and expenditures have by comparison been miniscule, and there is some relationship between this factor and influence on government policy. While we can see this more clearly on our side, there is a somewhat symmetric situation on the other side couched in a very different ideology. On some occasions when the arms control negotiators on the two sides have seemed to be approaching agreement, military influence has been able to pull the rug out from under them. In reviewing four examples of this we shall have a brief outline of nuclear arms limitation negotiations in the fifties and early sixties.

It should be appreciated that, at least in this country, each of the two groups has been seeking national security in its own way in a situation that is new in the nuclear age in which there is no longer any real security, since our being alive tomorrow depends on decisions of a foreign power. The military school of thought starts with the premise that the other side is apt to attack unless deterred by superior weaponry, and that the way to maintain superior weaponry is to develop and deploy new weapons systems more rapidly than the other side can. The arms control school of thought recognizes that the destructive power of nuclear weapons is so enormous that no national leaders could rationally start a nuclear war without vastly superior weaponry, no matter how great the national interest involved, that is, that rough equality on both sides is sufficient to deter, and that tensions would be reduced and war would be less likely to break out if rough equality could be permanently achieved through effective and mutually advantageous agreement. Neither school is proposing to "let down our guard."

Shortly after the advent of the H-bomb in 1953, after years of little change of position, there was a short period of active negotiation in the UN Disarmament Committee. The attempt was to balance Russian concessions in conventional arms, such as tanks, in which they were superior, against U.S. concessions in nuclear arms, where we held the lead. The Russians proposed percentage cuts in military manpower, leaving them in the lead in that department, and rejected continuous inspection of nuclear production facilities. There evolved a British-French-U.S. proposal placing military manpower under numerical ceilings (and leaving the three Western powers together almost twice as many men as the Russians) as a condition for complete elimination of nuclear weapons under what was considered adequate inspection and verification, to include inspection of nuclear power plants from which materials might be diverted for military use.

The Austrian peace treaty had shown a distinct softening of Russian foreign policy with the advent of Khrushchev. Then, on May 10, 1955, this suddenly appeared in the disarmament negotiations also. A new Russian proposal accepted the demands that up until then had been vigorously insisted upon by the West, including the sticky points of numerical ceilings and continuous inspection of facilities under control. This seemed to the Western diplomats to be a surprising and welcome breakthrough, an indication of genuine interest on the part of the Russians to conclude an agreement that would have avoided the nuclear arms race. Ambiguities and differences remained, but hopefully could have been resolved. The U.S. response was to withdraw the proposals it had made, saying, essentially, that the inspection we have been proposing won't do the job, and besides, we don't want to reduce military manpower. Incidentally, we proposed instead "open skies," unlimited photo reconnaissance. Rug pullout number one. Great disappointment in disarmament circles.

This retraction made it apparent that the American Department of Defense, which in practice has veto power over arms policy proposals, had not taken the opportunity of disarmament seriously enough to do its homework before the Soviets unexpectedly agreed to something considered sensible by our diplomats. The Defense Department did not want to give up its superiority of nuclear striking force with jet bombers, but had nevertheless permitted our negotiators to put on a good show by proposing nuclear disarmament. It was not until after this incident that the United States abandoned complete nuclear disarmament as its stated goal and, rather vaguely, substituted "partial" nuclear disarmament, or what has come to be called arms control.

Despite that setback, the negotiators kept on trying. More ambitious schemes having failed, negotiations turned to the possibility of banning the testing of nuclear weapons. American scientists had suggested this as a means of arms control, of slowing down nuclear weapons development. The Indian government had suggested it as a means of avoiding the radioactive fallout from the tests that was becoming biologically troublesome. At a UN disarmament meeting in London, the U.S. delegate (Harold Stassen) and the Soviet delegate (A. Zorin) seemed to be getting hopefully close to agreement on reasonable terms for a test ban when the chairman of the U.S. Atomic Energy Commission (Lewis Strauss) became concerned and suddenly, and dramatically, brought two leading weapons experts (Teller and Lawrence) to the White House to impress on President Eisenhower the importance of continuing tests to develop the next attractive bit of nuclear hardware, the so-called clean bomb. Stassen's mission was promptly withdrawn. Rug pullout number two.

In 1957, in the United States, it was shown that a nuclear explosion could be contained underground. Then verification of compliance with a nuclear test ban came to involve the technical problem of detecting underground explosions and of distinguishing the seismograph signals they made from those of earthquakes. In the summer of 1958, a conference of technical experts from East and West met at Geneva and came to an amicable agreement on the technical provisions of a proposed test ban treaty that would be reasonable from the point of view that it would be better to have a monitored test ban, leaving the technical possibility of cheating with a few relatively very small bomb tests, than to have testing continue without limit. This incensed some of the nuclear weapons people in the United States, who proceeded to invent and make much of the possibility that intermediate-size nuclear explosions could be "muffled" by carrying them out in very large underground cavities that turned out later to be impracticably large to construct. So the United States did not follow up diplomatically the Geneva recommendations. Rug pullout number three.

The subsequent test ban negotiations came to focus on how frequently "on-site inspections" would be permitted by outsiders on the territory of a nation to determine whether a suspicious seismic signal was caused by a bomb test or an earthquake. By 1960, a compromise on this question seemed to be in the making, and a "summit conference" of heads of state was set up at

Geneva, where, it was hoped, a final test ban agreement could be formalized. At just this time the Soviets shot down an American U-2 high-flying reconnaissance plane, proving that the United States had been practicing ''open skies'' on the sly. The Russian reaction was to cancel the summit conference. Rug pullout number four, this time from the Soviet side. . . .

The agreement by bilateral treaty against wide-area ABM deployment was the one firm and concrete accomplishment of the Strategic Arms Limitation Talks (SALT) between the two nuclear giants that took place from late 1969 to 1972. These talks were undertaken after it became apparent in the Eighteen-Nation Disarmament Commission that there were vital problems that could be settled only between the two nuclear giants, and after a rather long delay initiated by the United States near the end of the Johnson administration to await the advent of the Nixon administration. Though it is normal for diplomacy to move slowly, this delay, and the ensuing year and a half spent deciding what to negotiate about in the SALT talks, seemed almost designed to allow time for multiple, independently targetable reentry vehicles, MIRVs, to be developed beyond the point of no return. These permit a single ICBM to carry several—perhaps even ten—separate nuclear warheads, each aimed at a separate target. Permitting their further development and deployment was a tragic shortcoming of the SALT agreement (known as SALT I, as there is another series of talks just commencing). The tragedy is that MIRV was originally conceived and motivated as a way to penetrate a future large ABM deployment: it is an ''anti-anti.'' In negotiating away ABMs, the SALT I agreement removed the raison d'être of MIRVs but permits MIRVs—and MIRVs are very destabilizing. They permit one missile to kill many missiles in their emplacements. They make any disarmament agreement much more difficult because satellite reconnaissance these days can count an adversary's missiles, but not the number of warheads they carry. The SALT I agreement consisted of a treaty banning large-area ABM deployment and a five-year executive agreement essentially permitting each side approximately the arms deployments it had already scheduled, spelling them out with some regard for balance between the two sides. Besides the firm accomplishment of a limit on ABMs, the other less firm accomplishment of SALT I was an implicit recognition by both sides of the value of an approximate balance for the sake of stable deterrence. It left one side superior in some categories, the other in others, consistent with present realities, but with an overall rough parity. It controlled numbers of missiles without controlling numbers of warheads; it did nothing about bombers or about killer submarines that seek to destabilize the submarine-based deterrent; and it left the arms development race unhindered. It left scope for SALT II, more scope than hope, unless higher priority can be given to avoiding dangerous continuation and spreading of the arms race.

Freeze the Cruise

[1984]

The modern, technically advanced cruise missile is a very undesirable weapon, from both global and long-term national security points of view. In the short term, however, while the U.S. monopoly of it lasts, it is militarily desirable, so keenly coveted by all three of our armed services that it is perhaps politically naive to think of stopping it by public pressure.

Yet the administration should be made aware of what it is doing to the future in proceeding with massive deployment of the cruise missile. There is still a little time for a policy change.

The MX and Pershing II missiles which figure prominently in arms limitation discussions represent nothing new in principle—merely important modernization, in terms of greater accuracy, of the bulky ballistic type of missile that has long been the mainstay of deterrence. The modern cruise missile, even though generically related to the buzz bombs of World War II, represents an entirely new departure in nuclear delivery systems, so revolutionary that it ushers in a new and even more dangerous era of the nuclear age. The heightened danger comes both from its ease of concealment, ending the possibility of attaining verifiable arms control, and from its lower demands on massive construction, making it eventually more easily available to third world countries than are the huge ballistic missiles, despite the high technology of its guidance system.

As we prepare to cross a threshold into a new era, we should recall advice given in an attempt to avoid crossing an earlier and even more significant threshold, that from the atomic bomb into the hydrogen bomb era—from fission to fusion. . . .

[While testifying in the Oppenheimer hearings in 1954, Professor Vannevar Bush revealed that in 1952 he had urged the secretary of state not to permit the first hydrogen bomb test without first seeking an agreement with the Russians forbidding such tests, for that test ended the possibility of the only type of agreement that then seemed possible and "marked our entry into a very disagreeable type of world." His remarks are quoted in chapter 2 of this book, on page 42.]

The wisdom of this pioneer of the nuclear age seems directly applicable to the present situation. Then the hydrogen bomb had been invented but not tested. Now the cruise missile has been invented, tested, and deployed in a preliminary way. But the crucial fact is that it has not yet been massively deployed, and most significantly, not by the Soviets. Once this has been done, the extent and details of its distribution in various countries will be almost impossible to verify.

The situations then and now are similar in another way. Then we had atomic bombs, ranging in power up to a quarter of a megaton, on which to base deterrence. We did not really need the hydrogen bomb unless the Soviets also had it, in appreciable numbers. Today we have all three legs of the nuclear triad on which to base deterrence, and thus do not need the cruise missile.

If we simply refrain from deploying it, the Soviets would presumably go on to develop it, following our lead in development as usual. But there is good reason to belive that they would not deploy it, knowing that we would proceed to do so if they should. If we do deploy it extensively, the Soviets doubtlessly will too, and our advantage will have been temporary. Eventually, we will face the prospect of proliferation of those weapons to relatively less stable areas of the world.

Even without deployment of the ground-launched cruise missile in Europe, some three hundred have been deployed for air launch from B-52s, twenty per bomber. Further production for the air force is, however, being discontinued for a year or so, presumably in anticipation of a more competent model.

Eventually the number of nuclear cruise missiles will probably run into many thousands when production is resumed and extended to ground and sea launching. The navy has been considering installing up to three hundred per vessel. With this momentum, they could be stopped only by an early realization in Washington that their long-term threat far outweighs their short-term advantage.

Some form of verifiable arms control seems to be the only practical alternative to an indefinitely long arms race ending in nuclear war. It appears too unlikely that the race will simply peter out, each side concluding that enough is enough, and reducing arsenals to less dangerous levels without regard to what the other side is doing. The alternative to nuclear war, then, involves verification—and verification will be denied us by massive deployment of modern cruise missiles.

We have now developed, tested, and to some extent deployed cruise missiles—which the Soviets have not yet done. The Soviets know quite well what we have done, for they can see us, with our open society, better than we can see them. It is we who need verifiability.

If the Soviets produce a great number of these missiles, we will not be able to verify what they have done with all of them. We can verify with confidence only *whether* they have produced and deployed them in quantity. Massive production is much more visible than are individual cruise missiles, and restructuring substantial military units for their use is not apt to escape our intelligence-gathering capacity.

Until they have been produced in quantity, however, it is possible to reach a verifiable agreement not to do so. The best time to make the agreement is now, while the Soviets are not so far along and have the short-term incentive to stop our momentum in addition to the long-term incentive which we share.

While a cruise freeze would be included in a general freeze on nuclear weapons production and deployment, the cruise freeze should be considered separately because of the different incentives and problems of negotiation both in Washington and with Moscow. Possibly, a decision could be made to freeze cruise missiles while the debate on the MX and other MIRVed missiles continued. On another level, in terms of the economics and politics of resisting industrial demands for weapons procurement, it might be more practicable to decide on one item of restraint at a time. Cruise missiles should be considered the most urgent.

We need the cruise freeze to keep open the possibility of stopping and reversing the arms race. We need the general freeze to exploit that possibility now; to stop the arms race by stopping, rather than by saying that we intend to stop after more arming. We need to pursue them separately, lest difficulties in achieving the general freeze should rob us of the cruise freeze and foreclose future possibilities.

The requirements for some degree of verifiability are rather different for the cruise freeze than for the rest of the general freeze. For the cruise freeze, what is needed is only to know that the Soviets do not do on a large scale what they have not yet done. For the general freeze, we will want to know what they do with individual missiles of types they have already made, which could lead to lengthier debate. For either, it is important not to get bogged down in demands for unreasonably strict verifiability.

The freeze movement calls for the United States and the Soviet Union to adopt a "mutual freeze on the testing, production, and deployment of nuclear weapons and of missiles and new aircraft designed primarily to deliver nuclear weapons" as an "essential, verifiable first step toward lessening the risk of nuclear war and reducing nuclear arsenals." A strict interpretation of "verifiable" would be appropriate only if a freeze were to serve as the sole limitation of the arms race for the long term. But the freeze is intended mainly as a temporary expedient, a bridge toward more formal agreements containing explicit provisions for verifiability. Thus, within the concept of the freeze strict verifiability is not needed. The approximate verifiability available through the usual national means of detection should suffice.

It is clear that we will not enter into an agreement with the Soviets that involves trusting them in military matters except when they are acting in their own best interest. A freeze agreement would be in the best interest of both parties. And it would remain in the best interest of the Soviet Union not to risk getting caught violating it. Beyond the short term some modicum of verifiability is needed—so that there would be a substantial risk of detection in a major violation, and preferably enough to generate some risk in less serious violations. To demand more of verifiability would be to buy an unnecessary margin of safety at the cost of probably failing to achieve a freeze agreement soon.

The degree of reliability we require of verification procedures depends on how we perceive the practical meaning of the strategic balance. In the early sixties we failed to achieve the benefits of a comprehensive test ban when negotiations stalled over the difference between allowing two or four on-site inspections per year, to check on distinctions between earthquakes and underground nuclear bomb tests. This insistence on a detail slightly affecting the reliability of verification procedures made sense only if one assumed that a few relatively small clandestine tests might upset the strategic balance. Yet each side had thousands of hydrogen bombs—the United States many more than the Soviets.

Even when we had four times as many deliverable strategic warheads as the Soviets had, the deterrent balance was stable. We, of course, were not aggressively inclined, but if we had been we would have been deterred from attacking by their many fewer missiles. This illustrates nicely how unessential it is that the balance be nearly exact in numbers.

A freeze, to be effective, must be a quick freeze. That means not delaying long over demands for too strict verifiability. The primary aim of the freeze movement remains: "Freeze nuclear weapons production to stop the arms race." But a pressing subsidiary aim should be: "At least freeze the cruise missile."

7 Personalities

Admiral Lewis L. Strauss, a man of considerable personal charm, erstwhile banking wizard become wartime desk admiral and later dominant member and chairman of the Atomic Energy Commission (AEC), seemed to many of us in the scientific community to be unscrupulously cunning in his domineering way of getting things done. Furthermore, we disagreed strongly with most of his policies. He seemed to be the epitome and the source of much that was wrong with government nuclear policy in the early fifties. He was extremely powerful in determining policy. He wore two hats, as chairman of the AEC and as President Eisenhower's special adviser on scientific matters.

Scientists were annoyed that all the information reaching the president on nuclear matters was filtered through Mr. Strauss. His most distressing affront to the scientific community was his viciously vindictive treatment of one of the greatest of our number, Robert Oppenheimer, who was something of a hero among us not only for his scientific prowess in peacetime but for his brilliant leadership in the wartime development of the atomic bomb. His real transgression seems to have been that he was too brilliant in opposing and showing up Strauss in a conference concerning the shipment of radioactive isotopes to Norway. A hearing was held within the AEC to determine whether Oppenheimer should be branded as disloyal by being stripped of his security clearance because of youthful questionable interests that had long since been cleared up when he was appointed head of the Los Alamos bomb project. Strauss trickily blamed the instigation of the hearing on a letter from a minor official making the accusation, but it was Strauss himself as AEC chairman who transformed what might have been a routine hearing into a veritable inquisition in the spirit of McCarthyism by appointing a famous and un-scrupulous trial lawyer as prosecutor.

We may have hoped we were rid of Admiral Strauss in government when his term as AEC chairman expired, but he was then appointed a member of the president's cabinet as secretary of commerce, subject to Senate confirmation.

A big event of the year for most active physicists is the Washington meeting of the American Physical Society in about the last week of April, cherry blossom time. By chance or design the Strauss confirmation hearings before the Senate Commerce Committee were scheduled for that same week in 1959. By chance also that was the week I began my year as chairman of the Federation of American Scientists (FAS). Some days before the meeting I had a phone call from the secretary of the Commerce Committee inviting and urging me to appear at the hearings and give my views on the appointment. The committee was being fair in seeking contrary opinions. I understood that six other scientists were being invited so it seemed no big deal involving just a short time taken out from physics to make a few remarks, since I was to be in the city anyway, and I accepted. When I reached Washington I was somewhat dismayed to learn that all but two of us had chickened out; David Hill, a former chairman of the FAS, and I were to represent the scientific community.

On the train to Washington and in my hotel room I wrote a prepared statement on my portable typewriter. For years I had been feeling that one principal roadblock to achieving meaningful negotiations on nuclear arms limitation was the fact that most senators are too busy with local matters to pay any attention to the problem. I had been trying to promote interest in the subject. It occurred to me that here was a chance to reach some real live senators as a captive audience and that the subject was related to the prospect of having Strauss in the cabinet because he had been such a tricky opponent of arms limitation. It was an opportunity to expose the senators once again to the idea that national security would be improved by trying sincerely to negotiate mutual restraint with the Soviets while continuing the nuclear buildup rather than just to trust the buildup, what I then called a two-track approach to security rather than the one-track approach promoted by Strauss. The result was a hybrid statement made up of the sins of Strauss and the virtues of negotiating nuclear restraint. The evening before the hearing I checked it over with our FAS legal counsel, Dan Singer, who recognized it as unconventional for Washington and pretty strong stuff, but he approved of it in general.

There was a contrast as I ambled along a marbled hall of the Senate Office Building looking for the hearing room and the admiral marched briskly up to it flanked by an aide at either elbow, and thus it was also in the hearing. I was to have given my testimony early in the morning session so that I could have gotten back to physics but it was approaching lunchtime when I was called on. As is my usual custom, I spoke impromptu following the ideas of the prepared text, rather than reading it. Including questions from senators largely about me and my background, it was a rather lengthy presentation. At the end of it the chairman declared it was lunchtime and asked me if I could return the next morning for further discussion. But he did not adjourn the meeting before Senator Scott had interjected remarks indicating that he had it in for me and would rake me over the coals the next day. With the uncertainty of that threat hanging over me, it was impossible to concentrate on physics papers that afternoon or indeed to sleep a wink that night. The hearing was reported in the evening papers and as I chaired the meeting of the FAS council that evening I was asked to tell about it. As I expressed some misgivings, someone cheerfully observed, ''Never fear, we have two Davids and they have only one Goliath.'' In the end the side with the two Davids won by a devious and strange route.

The most significant part of my grilling the next day was that some of the questions Senator Scott put to me indicated clearly that he had had access to my strictly confidential AEC security clearance file. It was clear to Senator Clinton Anderson that he could have obtained it only from Admiral Strauss and that this was yet another of the admiral's dirty tricks. This angered Senator Anderson. He was in a distant way a friend of mine, knew me quite well enough to know that the insinuations were unfounded, and disliked the way they tried to smear an innocent scientist inept in the ways of Washington. Senator Anderson, chairman of the Joint Committee on Atomic Energy and thus the top nuclear man in Congress as a result of the bomb having been invented in his New Mexico, had long been at odds with Strauss as the top nuclear man in the administration. There was plenty of animosity between them but the senator had decided not to enter the fray and waste any of his political capital on the confirmation hearing because such hearings are routine and confirmation is almost a foregone conclusion. But my ineptitude in innocently lecturing to the senators on an unpopular subject and calling down Senator Scott's wrath upon me changed all that. Senator Anderson was well prepared and brought in his big guns and won the battle. Confirmation was in this rare case denied.

As I left the hearing after my grilling, the secretary of the committee asked me for a copy of my prepared statement. It being before the days of Xerox, I had only the one I had typed and let him have it on the understanding that it would appear in the transcript so I would later have a copy. What appeared instead was my oral presentation, which is all that is left to be included here.

Leo Szilard was a wonderful man at the opposite end of the spectrum of concern. He had a gentler charm, and an almost cherublike quizzical smile accompanying frequent rather extreme, partially tongue-in-cheek remarks. Of the four brilliant Hungarian-born scientists who had profound influence on U.S. nuclear affairs—Szilard, Teller, von Neuman, and Wigner—he was in later years the only dove. In his personal life he was a lonely figure, not married until rather late in life and then to a fine and dedicated woman whose medical profession kept her in St. Louis while he was at the University of Chicago. There I had the pleasure of lunching or dining with him occasionally at the Quadrangle Club where he lived on campus. Leo and Gertrude seldom lived in the same city until his last years of ill health, mostly in Washington, after his self-directed recovery from a bout with cancer.

He usually managed to live where he could best do what he wanted to do: in the thirties in England where he could ease the exodus of scientists from Hitler's Germany; at the end of that decade in New York where he could collaborate with Fermi on experiments leading toward the achievement by nuclear fission of the chain reaction that he had invented years earlier; in the early postwar years in Chicago where he pursued biophysics more than physics; and in the last years of his life in Washington where he could influence the senators' attitudes toward the arms race.

In tribute to his memory the Forum of the American Physical Society in 1974 established an annual award known as the ''Leo Szilard Award for Physics in the Public Interest.'' Perhaps largely in recognition of writings such as the ones in this collection I had the honor of being its first recipient. The occasion of making the award was once again during the spring meeting of the Physical Society in Washington and in accepting I again spoke informally without notes (this time because I left my manuscript in a taxi a couple of hours before the speech!) but the substance of my remarks is contained in the second paper of this chapter.

Nomination of Lewis L. Strauss

[1959]

[Hearings before the Committee on Interstate and Foreign Commerce, United States Senate, Eighty-Sixth Congress, March 17–May 14, 1959.]

Statement of Dr. David R. Inglis, Chairman, Federation of American Scientists

Dr. Inglis: I have a statement to which I shall stick rather closely, but not quite exactly. I have been requested by this committee to come and express my views relevant to the question before it.

Senator Thurmond: Mr. Chairman, could I ask this question of the witness at this point. Did you voluntarily come here or did someone ask you to come; and if so, who asked you to come?

Dr. Inglis: Your secretary, Mr. Cox, called me on the phone, I think it was last Thursday, and told me he was very anxious to have the views of the scientists presented and would I be willing to cooperate with the committee and come. He mentioned the possibility of the convenience of a subpoena, which I waived. . . .

I personally find it distasteful to be in a business that might seem like personal recrimination, but under the circumstances I feel it my duty to cooperate, even though it is an unwelcome task, because it must take the form of criticizing an important individual who seeks assiduously to serve his country as he thinks best, and who has the confidence of President Eisenhower.

I cannot hope to compete with lawyers accustomed to the ways of Washington in the completeness of my brief, especially on short notice, but I shall reply to the committee's request by trying to convey my undocumented opinions after observing for some years how Mr. Strauss has influenced the development of attitudes and policies in the United States. I have two major points to make, of which I elect to discuss first and in greater detail the one which I suspect may not be emphasized in most other statements before this committee. It concerns the part of Mr. Strauss in frustrating a broader attack on the problem of surviving the atomic crisis in national security.

Though we frequently forget it in our daily activities, I believe we must all agree when we really think about it that our national security, now and in prospect in the years ahead, is in pretty bad shape. This is in spite of our strength in strategic and tactical nuclear weapons and the means to deliver them, in spite of our system of alliances and our defensive warning networks, in spite of our economic strength and our scholarly prowess in science and many arts.

Our ability to deter Russia and potentially other countries from expansionist activities rests in making threats of virtual annihilation which, if carried out, would mean our own virtual annihilation. We feel confidence, and perhaps more confidence now than a few weeks ago, but not quite with certainty, that we shall be able to get by the Berlin crisis by these means. However, even if the contest should for many years remain one between us and the British as atomic nations on one side and the Soviets on the other, we could not feel confidence in getting by many crises, one after another, without touching off the war that nobody wants. The threat would be ineffective as deterrence if it were not meant that seriously.

Our situation is, however, even worse than that because there will fairly soon be many nations able to make nuclear weapons and to threaten disastrous war, and then the problem of avoiding accidental war will be truly a nightmare. We are marching almost blindly into an intolerably dangerous situation, and our chances for surviving as a recognizable nation through many decades seem rather small.

Our national guilt—if I may put it that way—for getting ourselves and the world into this unhappy fix without looking really seriously for acceptable alternatives is due in no small measure, in my opinion, to the narrow dedication of Mr. Strauss to the single-track approach of modern weaponry with no toleration for negotiations as a parallel track toward future security.

I do not mean to imply that we have not negotiated nor that the Russians are not at least equally guilty, but I do assert that we have not negotiated nearly as seriously as the situation has warranted. Our proposals on arms limitation have been hastily improvised, and they have been halfhearted in the sense that, in many instances, the administration team could not agree on any proposal that the Russians could reasonably be expected to accept.

We have not made the negotiation of worldwide limitation on the development of atomic arms a primary aim of our foreign policy. There have been isolated men high in our government who have tried to move in this direction, men like Robert Bowie and Harold Stassen. I perhaps shouldn't have omitted to emphasize here President Eisenhower's frequent expressions of extreme interest in this direction, for his very real efforts.

But, referring back to the men lower than President Eisenhower, they have been frustrated partly by opposition from the armed services, each specialized in and properly dedicated to its own task, but jealous of its share of the budget and accustomed to opposing any influences that might be suspected of interfering therewith. But they have been opposed, I think, even more effectively by one man with two jobs, Admiral Strauss.

The world crisis we are dealing with is primarily an atomic crisis, and in meeting it President Eisenhower has been dependent most heavily on advice on atomic matters. He made Mr. Strauss the head of the AEC, in recognition not only of his great personal charm, an example of which we have seen here this morning, but also of his effectiveness in getting things done and the importance of his dedication to modern atomic weaponry.

I presume the president had in mind keeping decisions simple within his team when he further selected Mr. Strauss as his special adviser on atomic matters. I doubt that the president fully realized to what extent he was cutting himself off from a balanced view in confining himself to a single channel of highly opinionated information and judgment on the problems of atomic energy as related to diplomacy. I believe he underestimated Mr. Strauss's ability to shatter opposing views by tactics which seem to me not always strictly ethical.

In speaking of negotiation, I do not mean to imply, either, that it is easy to negotiate with Russia—we all know some of the difficulties—nor that the most carefully prepared and earnestly motivated negotiations would with certainty have ended in useful agreements. I do mean that we and the Russians have very strong mutual interests in trying to avoid a war of annihilation and, for example, that we should both want very much right now to keep the nuclear club small, and that we have not as a nation sincerely put our shoulders to the job of trying to take advantage of these mutual interests.

As an example of the degree to which we have not even been really alive to the idea of negotiation I want to mention one isolated instance, the fact that in 1955 Premier Bulganin, who was then a top man in the USSR, suggested in a letter to President Eisenhower the setting up of ground inspection posts at important places throughout both our countries, as one means of suppressing surprise attack. To this small extent, he was offering to open up the Iron Curtain.

Instead of simply accepting this offer, or even exploring how much observation equipment could be put in the control posts, what we did was in effect simply turn it down, by attaching unacceptable strings. Even prior to this there were people who urged that we should stop nuclear tests as a means of controlling the development of nuclear weapons, but not until this last year have some of our policymakers come to realize how important it could have been if we had then set up ground inspection posts.

The reason our negotiators at that time were not given more scope was that Mr. Strauss and his close associates promoted their belief that a one-track atomic policy would keep us safe, or, more specifically, that we are inherently so much better than the Russians at research and development that in a free race without any limiting ground rules we can keep ahead forever.

I apologize for oversimplifying when I say that this belief ignored the long-recognized tendency toward saturation in nuclear weapons capabilities, the conditions wherein each nation has the capability of destroying essential enemy targets several times over, and neither has an advantage over the other except in the finer points of suddenness of surprise delivery.

We are now seeing an approach to this condition, along with the beginning of its many-nation aspects, and have little reason to be happy with it.

I sincerely believe that we would quite probably be in a better position today and face a much better situation in the future—and in any event would be no worse off—if we had tried much harder in the past, perhaps starting with a large-scale research project to discover and exploit all the possibilities of increasing national security through negotiations. Such an effort need not and should not have involved one whit of reduction of our buildup of weapons systems except as we might be able to elicit matching and verifiable concessions from the Russians and eventually from other potential nuclear nations. This second track would not interfere with the first. Two tracks would be better than one in trying to avoid a collision.

In opposition to such negotiation, one of the most effective methods of Mr. Strauss and his associates has been to point with enthusiasm to the improvements which we expect to make if we continue with unlimited development, refusing to recognize that the advantage thereby gained will be offset, and perhaps more than offset, by the improvements that our competitors in Russia will gain from their tests if they, too, go on with unlimited development. This is another aspect of his one-track approach.

The short-range quest for security only in the next very few years means forgetting that after some years of unlimited competition both sides will be trying to deter one another with much more refined weapons which will make possible more sudden and insidious surprise attack, making accidental all-out war more difficult to avoid, and making it technically still more difficult to bring weapons under some sort of effective international control if this should later become politically feasible.

There is now not much time to try by agreement to keep the nuclear club small, as a first step which would be valuable in itself and might lead to more extensive arrangements for security through arms control. But there is still a little time, and it is important that it not be wasted. The decisions of Mr. Eisenhower's administrative team will be crucial. It is already loaded with representatives of the armed services who have their special interests. That the chairman of the AEC and the president's special adviser on atomic matters should also have been opposed to negotiation has been unfortunate. I believe that the Senate should sense the danger and do its legal and appropriate part in preventing the president from, perhaps unwittingly, further loading his cabinet against a two-track policy by the new Strauss appointment to his cabinet.

My argument until now has depended on the proposition that we might with enough thought and united effort be able to convince the Russians to live up to an agreement which is to the best interest of both of us and will increase our chances of avoiding mutual disaster. Some senators will remain dubious about that, I am sure. So far as they can get time from pressing daily problems, I commend to them further thought on where we go from here, with many nuclear nations coming up.

As outside reading on the sort of thing we might negotiate, I suggest, for example, an article by Richard Leghorn in the *Bulletin of the Atomic Scientists,* for June 1956. There is another which I wrote in the same issue, and another in June 1959. I want also to insert a plug for Senate Resolution 96, which gives senatorial blessing to the negotiations at Geneva. Our approach there is all to the good, even if belated and perhaps not quite flexible enough.

My further citation of another very great and very detrimental influence of Mr. Strauss has nothing to do with negotiation which some of you may still consider a bit hypothetical. It has to do with our next generation of scientists. How do we get really good new American scientists anyway? The first hurdle they have to pass, unless they are very lucky, is that they have to be educated with a great crowd of other children who don't care as much as they do about learning, because we educate a lot more people here in America than in Europe, for example, and don't make enough special provisions for gifted students.

There are other hurdles in college, but another hurdle is that our really bright student sees quite clearly that he has a good chance of getting rich if instead of science he chooses a money-making career in our fabulous system of free enterprise. This is one respect in which it is very different in Russia. There, if he can pass the exams, he knows he'll be making top money as a scientist.

This would seem to be an advantage for the Russians in getting scientists. However, there is an advantage much more than capable of making up for that, which we have at times enjoyed and should bountifully enjoy. This is the advantage of spontaneous freedom and complete independence of thought, which appeals strongly to men of great breadth of imagination.

Up until now we have done much better than the Russians in fundamental research, and I believe it has been largely because of this background, although the Soviet economic troubles of the thirties and the great strength we have received from European talent and education have also had much to do with it. Complete independence of thought and judgment for scholars is a great national asset, a bulwark of strength for the future, and Mr. Strauss has deliberately dragged it in the dirt before the very eyes of the budding scientists of the next generation.

As far as I can learn, the influx of students of really high ability and imagination into the ranks of science today is dwindling when it should be rapidly growing if we are to take full advantage of the boundless possibilities of the future. This I attribute mostly to the anti-intellectualism which, in particular against science, has been enormously stimulated by the debasing action of Mr. Strauss.

You all know of the most famous example. The record is long and even tedious, although packed with human emotion and pathos. There is pathos in the fact that the long and very intimate record, taken under assurances of confidence, was blatantly published by Mr. Strauss.

The tone of the fact-finding hearing was changed to the tone of a trial when Mr. Strauss chose as "prosecuting attorney" Mr. Robb, who had similarly served the late Senator McCarthy. The board found loyalty but condemned for opinion. Mr. Strauss disagreed, sidestepped condemning for opinion but instead condemned for an ancient lie to cover up a friend, a fact long ago confessed and cleared up with General Groves early in the course of Oppenheimer's magnificent, brilliant, and unsurpassed wartime service to the nation.

The condemnation seems to me, and to others, to have been largely a matter of Mr. Strauss's personal vindictiveness, arising at first from an occasion in which Oppenheimer crossed Strauss in the matter of sending useful and completely nonmilitary isotopes abroad, about the technical side of which Oppenheimer naturally knew more than Strauss.

Here again I oversimplify a complex matter. After observing the proceedings this morning, I see that this brevity leaves me wide open to rebuttal and I can expand on it, if need be; but I have tried to give some brief indication of the basis for my opinion that it was Mr. Strauss who was most severely at fault. It was he who, partly out of spite, has alienated the most brilliant and independently thinking of the youngsters against science and has alienated large numbers of active and distinguished scientists, and it is he who, because of "substantial defects of character"—as he said—is unfit to serve on the president's cabinet, or to head a department in which he again would be administering affairs of science.

This ends my written and prepared statement. I have some further remarks which I would like to make, if I am not overstepping the bounds of time.

> *The Chairman:* No. we will be glad to hear any further remarks.
> *Dr. Inglis:* I would like to give some idea of my impressions of scientific opinions on this matter. I realize that they are not unanimous.

I should remark that most scientists are really engaged practically all their time and highly devoted to research and teaching and this leaves little time for political interests. Yet I feel that most scientists, even those without very strong political interests, have a strong distaste for Mr. Strauss and for his methods.

We scientists are far from Washington and far from the rumors about the infighting in Washington. I suppose you receive more of them than we do. I believe that this general distaste that scientists have arises rather largely from Mr. Strauss's role in the extension of security in science beyond what we consider its reasonable realm of applicability. In his first term in service in the Atomic Energy Commission, he served the function of rather the watchdog of security.

Now, there are some matters in the Atomic Energy Commission's work in particular, and in science in general, which we all agree are very properly matters for strict security measures; but there are other matters in fundamental science where we feel it is entirely wrong that security measures should be applied.

It has, in leaving the wartime situation, been a real problem to sort out these two parts of the interests of the Atomic Energy Commission and of science. This sorting out has progressed and has now gone to very satisfactory lengths, but it was very slow. There was a long time when scientists felt a very strong animosity to Mr. Strauss in his role of hindering and slowing up this process. . . .

He mentioned this morning that it was his idea to have the Geneva Conference, the International Conference on Atomic Energy. I agree, it was his idea, but I feel that this was, in a way, a long-belated recognition of an inevitable trend and a superb act of showmanship, after having been wrong about security for many, many years—of deciding now that we are releasing these security measures, we will make a fine show of it. We did. I agree that Mr. Strauss had some part in this at the time, but I think the scientists perhaps may also claim some credit for the success of the show. . . .

Well, to summarize, I would like to say that I feel that the narrow approach of Mr. Strauss to one very necessary part of our foreign posture, namely, to our position of strength, that the narrowness of this approach has long delayed our proper attention to the other part of an old slogan, that we would negotiate from positions of strength, and has thereby done a disservice to our proper attack in trying to solve the most important problem of the world, the problem on which our very survival most depends, in a very unfortunate way.

Although he has made as an administrator, an able administrator, contributions in a narrow sense to the material needs of science—he has brought to you senators a proper analysis of the material, the financial needs of science, and helped to make provision for them, as I think any other able administrator in his place could have done, given, as he mentioned this morning, the efficient operation of the Atomic Energy Commission, which he inherited—in spite of these positive contributions, it seems to me he has made much more than compensating negative contributions in detriment to the spirit of science, the spirit of freedom, and the possibility of recruiting bright new minds to the strength of our science in the future.

These are some of the reasons why I, and I think a considerable majority of American scientists, feel opposition to the appointment of the nominee whose confirmation is before you.

The Sweet Voice of Reason

[1974]

Leo Szilard was a remarkable thinker. His inventive mind delved deep into more than one science and ranged over unique approaches to political action. His concern for the misuse of the nuclear chain reaction by society covered a long span of his professional life. In this he was at first far ahead of his time, for he invented the chain reaction shortly after the discovery of the neutron in 1932. A neutron-induced nuclear process yielding both neutrons and energy was a mere figment of his imagination, long before the discovery of fission in 1938. His concern commenced with the invention. Hypothetical though it was, he patented the idea and consigned the patent to the British Admiralty where it could rest in secrecy and presumably do no harm.

When fission did come along and concern arose that Nazi Germany in World War II might be the first to exploit nuclear explosives, it was Szilard, along with Eugene Wigner, who had the idea of getting Einstein to sign a letter to President Roosevelt, embarking the U.S. government on the long road to nuclear weapons and eventually to the promotion of industrial nuclear power. Having invented the chain reaction and having been among the first to promote its development into a terrible weapon, after the defeat of Germany Szilard devoted much of the rest of his life to reducing the likelihood that it would be used as a weapon of war. He was among the group of scientists who virtually gave birth to the AEC by promoting the MacMahon Bill. When subsequently the production of weapons seemed to be going to unnecessary extremes, he devised his own special ways to "bring the sweet voice of reason to Washington," as he was fond of saying.

He felt his job was to think, and when doing experimental science with others, even with so prestigious a collaborator as Enrico Fermi, he left it to others to sully their hands. He must have had in mind his own act as pure thinker when he assigned this role to the handless dolphins in his whimsical yet profound little tale of imaginative international politics for the nuclear age, *The Voice of the Dolphins*.

A corollary of thinking was talking to the right people, and he was energetic in getting to the right place to instill his ideas where they might best count. Finding freedom to travel from his professorship at the University of Chicago, he spent much of the last part of his life in Washington, the part he had gained by overcoming cancer with the help of his own studious efforts.

Szilard saw the U.S. Congress as a crucial center of power for worldwide good or evil, rendered partially ineffective by the competing influences of money from vested interests. He believed that decisions of Congress could reduce the likelihood of nuclear war through moderation of the arms race. His approach to bringing the voice of reason to Washington was twofold. First, thoughtful scholars with legitimate social concerns should talk to members of Congress, and he set a good example. Second, money talks, and socially concerned citizens should make the relatively small amounts of money they could raise talk most effectively by selectively supporting liberal candidates in election contests that promised to be close, where a judicious concentration of funds could be most effective. He founded the Council for a Livable World to this end and it still carries on in the same spirit, having really helped to increase the number of members of Congress favorable to a policy of moderation.

How close efforts of this sort came to success was dramatized by the famous fifty-fifty Senate vote on deleting funds for the antiballistic missile system. Here one more vote might have drastically changed the course of arms control. While the SALT talks ultimately did eliminate large ABM systems that would have provided a raison d'être for MIRVs, they were not eliminated until after the Department of Defense had successfully developed MIRVs, presumably against the prospect of ABMs, while the preliminaries to the talks dragged on and on.

It is sometimes questioned whether it is worthwhile applying reason to the problems of regulating the arms race when the actual decisions are political. But the concerns of contention in politics have their origins in more abstract discussion. Today's reasoned argument becomes tomorrow's political force. Witness, for example, the idealistic idea of a nuclear test ban. It was discussed in this country among scholars as an element of arms control, and championed by Adlai Stevenson as a way to avoid radioactive fallout. It was first denounced by President Dwight Eisenhower in his presidential election campaign, and then espoused by Eisenhower. Later, under President John Kennedy, it was partially adopted in a limited test ban as an international political reality. Unfortunately now, eleven years later, a complete test ban has not yet been achieved.

Two to Make a Race

One may wonder whether an improved negotiating position in matters of arms control on our side would have been reciprocated by accommodations from the other side. It takes two to make a race and it takes two to stop an arms race. Throughout most of the course of the nuclear arms race the United States has been far in the lead. If the method of stopping the arms race were to have been a sequence of unilateral steps of restraint, unilaterally reciprocated by the other side, it would have been up to us to make the first move—but this we did not choose to do. The hope instead was for mutually advantageous and formally accepted steps of arms limitation, and eventual reduction, monitored to a reasonable degree. For such an agreement it is necessary that those in power on both sides understand the possibility and the mutual advantage of reducing the likelihood of the outbreak of a nuclear war that nobody wants.

Back in the mid-fifties when ideas of this sort were circulating among concerned scholars in the West, there was no evidence of similar thought in the USSR. The breakdown of most of the barriers of secrecy in science in 1954 made possible contacts between scientists of the East and West. It was urged by Einstein, Bertrand Russell, and others that scholars of East and West should be brought together in discussion of such matters, and Szilard was prominent among those who participated in the subsequent Pugwash Conferences on Science and World Affairs. The motivation here among Western scholars was not only to develop ideas about arms control and disarmament but also to plant the seed of such ideas within the USSR. In this effort, as elsewhere, Szilard had his own personal style. I remember how at a meeting in Moscow in 1960, while the rest of us methodically attended the scheduled discussions in the modest hall known as the "house of peace," Leo was much of the time off elsewhere in the city making his own contacts.

Since academic ideas do sometimes diffuse upward in the realm of politics, it seemed particularly important to see to it through such conferences that Soviet scholars should be pursuing thoughts similar to those circulating among Western scholars in matters of arms control, so the upward diffusion could take place simultaneously on both sides and eventually political leaders on both sides might talk the same language and have similar concerns.

Since the first of the Pugwash conferences in 1957 East-West discussions of these matters between scientists have continued. There is evidence that some of the ideas have diffused upward, the opinions of the groups favoring arms moderation within the governments of the two major nuclear powers have to some extent converged. Despite great differences in ideology and some remaining differences in outlook on arms control matters, there is a degree of symmetry.

But this has not been enough. The arms race continues. The trouble is that the symmetry goes further, for on each side there are the military establishments and supporters promoting continued armament. The slow progress—the almost total lack of progress—in arms control attests to the fact that at least on one side, and probably on both, the pro–arms race faction has held the upper hand in determining national policy.

There is positive feedback in this situation. If the hard-liners on one side succeed in dominating the negotiations and imposing unattractive conditions, this helps the hard-liners on the other side retain the upper hand. Thus it makes sense to try to alter the situation on our side.

In our relatively open society, it is easy to observe the extent to which the hard-liners do have the upper hand on our side. One somewhat distorted measure of it is the appropriations for the Department of Defense versus the Arms Control and Disarmament Agency—$80 billion versus $9 million.

The overwhelming influence of the hard-line position in the United States arises from a feedback loop that is probably much more effective in our free enterprise economy than is any counterpart there may be on the other side. It has gone by the name given by President Eisenhower, the military-industrial complex. But the loop is larger; it is a military-industrial-labor-legislature feedback loop. Physicists are included in each of the first three elements of the loop, but unfortunately only minimally in the fourth, Congress.

The effectiveness of this loop stems from the great triumphs of applied science left free to proliferate without due consideration of the consequences or introduction of compensating political mechanisms. Industrial processes have become so efficient because of the development of technology that society is freed of much of its former drudging human labor, and the natural consequence is widespread unemployment. This should ideally provide opportunity for even greater improvement of the human condition than we have seen, increased amenities and services, enriched uses of leisure and works of the mind, with readjusted distribution of abundant consumer goods based on less than a full work week. But this would require regulation or at least redirection of free enterprise. Instead we have relieved unemployment and provided profitable stimulus to the growth-oriented economy (and incidentally to science) by production of socially useless goods, notably military hardware. For continuation of this situation, labor joins the military and industry in bringing pressure to bear on Congress for bloated military appropriations. Proponents of arms control propose conversion of facilities and employment to production of needed socially useful goods, at least as an interim measure, but as yet to little avail.

The policy gulf between arms controllers and the supporters of the arms race arises from the efficiency of another product of science and technology, namely the awful destructiveness of the nuclear weapon. Reasonable extrapolation from the gruesome samples of Hiroshima and Nagasaki, and comparison with the prospective cost-effectiveness foreseen by those who started the great conventional wars of the past, can be seen to lead to the conclusion that the prospect of being on the receiving end of a hundred nuclear weapons, or even ten, should be enough to deter deliberate nuclear attack. Yet, conventional military wisdom teaches that one wins by being there first with the most, and military traditionalists insist on piling overkill upon overkill, not only by stockpiling thousands of deliverable nuclear weapons but also by pursuing every proposed improvement in their prospective effectiveness, regardless of the possibility of negotiating a mutual halt. This has been the situation all during the time when Leo Szilard was trying to bring the sweet voice of reason to Washington, and still is. The turnaround and approach to the low-level stabilized deterrent that he and others promoted in his lifetime seems as far away as ever.

There is a ray of hope in the fact that it is now almost thirty years since the nuclear weapon was used to end an otherwise conventional war and, despite all the conventional warring in the meantime, there has been no outbreak of nuclear war. This should not be allowed to give rise to complacency, as it seems to do, for thirty years is a small sample of human progress.

Until recently it seemed hopeful also that there had been no new entry into the nuclear club for a decade. Now that as impoverished a country as India has tested its first nuclear explosive, the hope has dimmed that there will not be several new nuclear weapons nations in the near future. This event emphasizes the folly of the nuclear nations in not having fulfilled with a sense of urgency their obligation under the Nonproliferation Treaty to pursue ''effective measures related to cessation of the nuclear arms race at an early date.'' They would do well to feel the pressure of events and mend their ways. As for India, this nation should be urged to declare a policy of stopping with that one shot, thereby belatedly setting a much needed example of restraint consistent with the spirit of India's great philosophers.

How generally various countries and even terrorist groups will have ready access to nuclear materials and nuclear weapons in the future will depend substantially also on how we develop our future energy sources, now so much under discussion. This is not the place to make a cost-benefit analysis of nuclear power production; but one of the severe costs is that generation of nuclear power involves the production of plutonium, nuclear bomb stuff and incidentally an easily dispersed, highly toxic material. There is already a lot of it locked up, but the proposed increasingly copious production of plutonium will make much more likely its diversion to undesirable channels. Nuclear power at home will provide plutonium for a potential black market. Export of reactors will put ever more countries, including small and unstable ones, into the plutonium production business. This is the thrust of worldwide competition in which we are leaders. We should be leaders in finding a better way, involving less risk of nuclear war.

There are promising alternatives to nuclear power production that may be made to look far off only because they have received ridiculously small exploratory funding as compared to the billions of dollars that have been poured by government and industry into nuclear power.

It would be much more in keeping with the needs of the less industrialized countries for us to export solar-related power technology than to export the plutonium production business. We are ignoring the risk of nuclear proliferation in our cost-benefit analysis when we fail to push alternative sources of power more vigorously than nuclear. That risk is one of several costs, and some view it as much too great a cost.

The administration bill before Congress proposes a greatly stepped up research and development budget for all power sources ($10 billion in five years), with most still going to nuclear. The proportion for all solar-related sources together has been stepped up to a meager 2 percent of the total. It is claimed that this is all that can be usefully devoted to this field as these sources are not expected to become economically important in this century, clearly a self-fulfilling prophecy in the interest of continued growth of the nuclear sector.

Applied science has placed unprecedented power in the hands of those who already have much and has increased their ability to acquire more power. This has led to exponential growth in many sectors, such as the doubling of per capita electric power consumption every ten years. This gives people great confidence in the ability of science to go on providing solutions as problems arise. It leaves them with little concern for what will happen as exponential growth exhausts the bounty of our wonderful green planet. It leaves them with little tolerance for restraint *now* to avoid the cataclysm at the end of a giant boom and bust.

Aside from the sheer beauty and satisfaction in the intellectual attainment of mastering many of the secrets of nature—a mastery in which the general populace too should be able to take pride—the greatest achievement of science is the richness of material life it has provided for so many people and the potentiality that some of these benefits may be extended to all. The threat of science is in its overuse. In meeting as physicists our responsibility for physics in the public interest, and for promoting the application of physics in the right places, we should with foresight and a combination of technical and political judgment seek to avoid excess in its use. In the domain of nuclear power, particularly as it relates to the arms race, we should continue to bring the voice of reason to Washington.

8 Energy

The problems posed by the two aspects of nuclear energy, its fast release in bombs and its slow release in reactors, are closely intertwined. Reactors make the stuff that bombs are made of, plutonium. Access to a complicated fuel-processing plant is needed before the material is ready for making bombs, but nevertheless having an industrial power reactor is an important step toward the capability of making nuclear weapons. If nuclear reactors were unique as a safe, dependable, and economical power source, as is officially presumed, there would be a legitimate desire to help third world countries acquire them to improve the lot of their teeming millions. Enhanced prosperity might even make some states less belligerent. Weighed against this presumed benefit is the increased likelihood of nuclear war as nuclear materials from power reactors in various countries get into the hands of irresponsible national leaders or even terrorist groups. There are other considerations in weighing the balance but in my judgment avoiding increased likelihood of nuclear war is by far the most important objective.

In the first of these papers, "Atomic Profits and the Question of Survival," we see the problem shaping up in the early years of the atomic age when atomic secrecy prevailed and when nuclear reactors were being developed under strict government auspices in the national laboratories with help under contract from a select group of industries that were let in on the technical secrets. The question then was whether to release the information and commercialize atomic power in the interests of more rapid progress, as indeed was done in 1954 under the slogan "Atoms for Peace." I then apparently shared the high hopes for safe and economical nuclear power, as I no longer do. I also had hopes for an arms control agreement based on accounting for all nuclear materials and saw secrecy as an impediment to determining how much had been produced. One sees here the premonitions of the arms race which now threatens us and the hopes for avoiding the commercial pressures that propel it, arising from a nuclear energy industry.

The case for exporting nuclear power plants as a way of relieving the suffering in underdeveloped countries was presented thoughtfully by Alvin Weinberg, director of Oak Ridge National Laboratory. He envisioned as an example a community established where the desert meets the sea, kept self-sufficient by production based on energy from a nuclear plant. Basic to the argument was the postulate that the population explosion is following its own timetable and will naturally level off only after a few more generations, there being nothing technology can do about it but try to meet its needs. It is this assumption that I challenged in the 1970 article "Nuclear Energy and the Malthusian Dilemma," as part of my long discourse with Dr. Weinberg that included a lecture he invited me to give to a staff meeting at the center for nuclear research that he heads. My message was, and still is this: the sooner nations and people perceive that the magic of science will not forever pull all needed miracles out of a hat, the sooner they will become serious about limiting birth rates. I still had hopes for breeder reactors for the long term but felt that the temporary use of uranium burners is adding to the illusion of future global plenty that encourages the population explosion.

I had long felt that the energy promise of the nucleus should be widely exploited only after agreements had been achieved to halt the arms race and the proliferation of nuclear weapons. The Nonproliferation Treaty and International Atomic Energy Agency (IAEA) safeguards against diversion of materials from reactors were steps in the right direction but woefully inadequate. As the number of incidents increased in which reactors went wildly out of control (making it seem only a matter of remarkably good luck that a bit more lack of control did not induce a burst of containment and a local radioactive calamity), I came gradually to believe that the inevitable inadequacy of commercial safety standards is an independent reason why we should stop building nuclear power plants. In the title of my 1973 book *Nuclear Energy: Its Physics and Its Social Challenge,* the challenge referred to was the challenge to reap the potential benefits of nuclear energy without unduly risking its awesome hazards, especially from bombs, but also from reactor accidents.

The potential of nuclear energy is attractive mainly because the coal, oil, and natural gas on which we depend for energy are a once-only boon from the geological past, more valuable for other needs than energy, being nonrenewable sources that will not last forever and that will rapidly become more difficult to extract from the earth. Reduction of our excessive energy demands through conservation could reduce the pressures for nuclear energy, but in the long term the development of alternative renewable energy sources is needed. I gradually came to appreciate that there are promising alternatives the development of which was being slowed down—even deliberately slowed down—by the priority being given to nuclear development both by government and industry. Nuclear proponents are ardent in denying the promise of alternatives. Thus promoting rapid development and exploitation of alternative sources seemed to me to be a practical way to fill a need and oppose excessive expansion of nuclear power. A bit of study of the alternatives led me to consider that wind power and ocean thermal power are the most promising, wind power being the one technically most ready to produce large amounts of commercial power soon and economically if promoted on a large enough scale. Led on by the enthusiasm of a colleague who is an expert wind power engineer, Professor William Heronemus, I wrote articles and a book on the subject, comparing the prospects of wind power with other energy options, including nuclear.

Most people think of windmills as those beside houses or Dutch canals, and it takes imagination to contemplate thousands of modern, megawatt-scale windmills on the windy western plains producing electric power on an adequate commercial scale and coupling them with the storage reservoirs of our great hydroelectric dams to supply steady power despite the fickleness of the wind. This aim is something that will probably be achieved sometime in the next century by successive waves of conservative investment in normal business practice but could be achieved much sooner if the government in the public interest would invest in it anywhere nearly as heavily as it has invested to speed the development of nuclear power. Since 1973 there has indeed been a federal wind power development program, but on a miniscule scale compared with the nuclear development program and always under the administration of the nuclear authorities who, with their excessively cautious approach, have seemed to want to demonstrate that wind power is for the next century and not a potential competitor of nuclear power now. This modest program and similar efforts abroad have slowly succeeded in improving the technology of large wind turbines (several megawatts) and now the time is more ripe than ever before for launching a deployment on a larger scale than commercial interests seem willing to undertake alone in competition with fuel-consuming energy sources that benefit from heavy federal subsidies. Some of the large wind turbines encounter difficulties such as short lifetime from metal fatigue, and more innovative engineering than the conventional approach by large companies may still be required to find the best designs. The technology of medium-sized windmills, in the hundred kilowatt range, has been more easily developed by small industries. Tax incentive programs together with state regulations for energy marketing have resulted in a proliferation of these smaller machines. Impressive arrays of them in windy passes in California are perhaps a small-scale foretaste of what might be done with larger turbines on the western Great Plains.

It is difficult to get large new enterprises started and difficult to stop them once they have become too large or have outlived their usefulness. It took an enormous initiative to achieve the first nuclear reaction, make the first atomic bomb, or develop a power reactor, and a lot of governmental arm-twisting of industry to get it into the nuclear power business. Now that the industry has demonstrated some weaknesses of nuclear power yet retains the momentum of vested interest, it seems to require more initiative than we are able to muster to replace it quickly with less dangerous power sources. There is an interesting parallel in this respect between nuclear power and the arms race. There, too, it took a lot of inventiveness to get the arms race started, and we seem to lack the will and initiative, as well as the perceptive judgment, to replace it with a less dangerous approach to national security through negotiated reciprocal restraint.

In my early articles and my book on wind power I made ballpark estimates indicating that very large scale wind power sources, when supplemented with partially underground pumped storage facilities to supply dependable continuous power, would cost no more than nuclear power from projected new plants. A typical case would be thousands of mass-produced megawatt-scale windmills on the windy western Great Plains sending power by long transmission lines to midwestern cities. Several groups of engineers in Europe and America, after completing their programs of testing large windmills generating electricity, had made estimates of what similar windmills would cost in quantity production. My estimates were based on translating their estimates into current dollars and comparing them with nuclear costs on the assumption that various incidentals such as the cost of investment money would be the same for both so the relative costs of the generating, storage, and transmission installations would be indicative of the relative costs of power. More recent experience with the cost of building the new mammoth machines by the industrial giants Boeing and Hamilton-Standard suggests that my estimates were optimistic. However, projected nuclear costs have skyrocketed, and the evidence still is that mass-produced big windmills should compete favorably with nuclear power. Those earliest estimates were based on production by smaller industries devoted more completely to the wind project, and the difference may arise from overhead costs and arbitrary work assignment practices in the large corporations.

The emphasis of my promotional efforts in the mid-seventies was on seeking federal support for independent entrepreneurs, able engineers with innovative ideas and ambitions to make large windmills. I was particularly enthusiastic about two imaginative projects, one on land and the other at sea. Charles Schachle in the state of Washington had devoted a lot of time and his own money to experimental developments and to a successful demonstration on a half-megawatt scale of a design and construction method that promised economies, a three-bladed windmill with a broad rotating tower and a hydraulic link for frequency control. He was well on his way to building a much larger version of it at a windy site in California under contract when he ran out of funds and had to sell out to a large company that assigned company engineers inexperienced with windmills to redesign it, eliminating some of its best innovative features, quadrupling its price, and making a fiasco of it. Another half-million dollars injected at the right time might have provided a useful test of this interesting variety of large windmill.

The other was the proposal of my colleague, Professor William Heronemus, to supply New England with its needed electric power by over ten thousand very large floating windmills exploiting the strong and relatively steady winds at sea some tens of miles offshore. To get started a demonstration was needed of at least one fairly large floating windmill, and Professor Heronemus stood ready to supervise such a project if one or a very few million dollars could be made available. When the administrators of the federal wind power project refused to touch anything so innovative Senator Kennedy got interested and introduced an amendment to an appropriations bill requiring the federal project, after a feasibility study that the senator presumed would be prompt, to submit a proposal for demonstrating at least one large offshore windmill. The deliberately evasive response to this mandate provides the clearest demonstration I have seen that the energy authorities have been sabotaging big wind power, avoiding exploring innovations that could help it become a major energy source in this century. The single half-million-dollar feasibility study contract was awarded not to anyone interested in such projects or with relevant nautical experience but to the nation's largest maker of nuclear power plants. A floating windmill of course has added costs for flotation not encountered on land, but there are compensating economies in the construction of an appropriately designed floating unit, as proposed by Heronemus, in addition to the great advantage of the strong offshore winds.

One economy is that a moored windmill would point into the wind the same way an anchored ship does and there is no need for the special machinery for this that is necessary on land. Another is that the floating unit could be completely assembled under factory conditions in a shipyard rather than in the field or at sea, and then towed to its mooring.

When the four-volume report of the feasibility study finally appeared after three years of delays, it completely ignored these economies. Indeed, it completely ignored all previous work and made no references to the literature on floating windmills. Instead it designed from scratch very expensive types of units, consisting of steady floating platforms simulating sites ashore (providing the "real estate," as the study said) on which to construct large windmills already designed just as they would be constructed on land, transporting parts by tugs rather than trucks. It is discouraging to think that this ridiculous report was taken at face value in Washington and there is little hope for aid from that quarter in proving that an appropriately designed floating offshore windmill can weather the storms out there as engineer and "old salt" Captain Heronemus believes it can. If the government, which has supported other power sources with which this must compete, will not undertake the first exploratory demonstration, it might be hoped that a commercial company would but none has, there being no prospect for profit until later stages. Thus this exciting possibility for bountiful future power remains unexplored by Yankee ingenuity.

The prospect for economical production of power from the wind to contribute importantly to national power requirements is better on a large scale, with thousands of large units commercially maintained in windy regions either on land or offshore, rather than on a small scale with millions of small units for individual homes. Power is needed for the cities as well as for farms and there are not millions of homes that combine a windy site with a mechanically inclined owner wanting to maintain a windmill. Down-to-earth types among environmentalists delight in "small is beautiful." That is to be encouraged as far as it goes for those who can achieve remote independence, but my emphasis has been on "large is effective" in meeting a national need and competing with nuclear power in an environmentally sound way.

Atomic Profits and the Question of Survival

[1953]

Decisions concerning possible changes in the Atomic Energy Act to permit less restricted industrial participation in the atomic production program should be made solely on the basis of their potential contributions to our national security and the general welfare of the people of our nation and the world. The profit motive may be used as a legitimate means toward this end, but must not be allowed through disproportionate influence of special interests to be made the primary goal of the legislative changes, the survival question being so much more important than the capital opportunities.

National security depends, in the short range, in part on the success of our present attempt to build up our military strength relative to that of the East, in which respect atomic weapons are of course vital, and in part on our success in maintaining unity within the Western world and alleviating the suffering of mankind in general, to which peaceful atomic developments or the prospect thereof may make substantial contributions. But our national security, along with the security of other nations of the world, depends in the long range on finding some way to end the accumulation of competing atomic stockpiles before they destroy us all. The probability that the armament race may be terminated without disaster depends on the chance that international agreement may be obtained on some form of armament limitations and control. All these aspects of the security problem would be affected, in different ways, by liberalization of industrial participation in the atomic production program.

The Short-Range Program and the Secrecy Question

A more commercial basis for atomic production can be justified only if it is found on technical analysis that such production is likely to increase or cheapen the supply of atomic explosives, or atomic fuels, or the application of atomic power, beyond what is attainable with the present AEC contractual arrangements, and thus help solve the security problem. The technical analysis of the prospective benefits, and indeed of the prospective profits and economic feasibility to atomic power in general, is greatly complicated by the scarcity of information we have at present. Drastic reduction of present measures and criteria for the security of information are difficult to justify adequately because of a widespread, and perhaps exaggerated, concern for the possible negative effect of the release of information on our short-range national security. It is really quite difficult to decide what sort of information, if released, may be expected to contribute substantially to the progress of the competing atomic buildup,[1] and the nontechnical public and its representatives quite understandably tend to be conservative on this point, since they cannot be informed on the exact nature of the questions involved. This is perhaps the most serious barrier to a completely rational decision on the prospective changes in the laws related to industrial atomic production. It is also a barrier in the more general problem of obtaining a public opinion sufficiently informed and sufficiently convinced of the real dangers to permit maximum efforts for security.

The hope in proposing a change is that the initiative of private enterprise, which has been a source of strength to America in other fields, may become operative in the atomic field as well. The extent to which this hope may be justified is in part hidden behind the wall of atomic secrecy. One glimpse into the question is provided by the reports of the industrial groups that were invited to look behind the wall of secrecy, to determine the extent to which their companies might be interested in investing in power development in the near future. The general consensus of these seems to be that little if any private capital would be forthcoming for developing power alone, and that a guaranteed price for plutonium is a prerequisite for the development of power as what we might realistically call a by-product ("co-product" is the usual term). There are quite naturally also requirements for government participation in the first phases of the industrial technical developments, to facilitate the transfer of present know-how from AEC to commercial groups. It seems clear that for the immediate future, at least, these groups are not willing to gamble a great deal on development in the independent manner characteristic of private enterprise in most other fields, but are willing to accept the results of government development (and atomic fuel manufacture) if they can get into the business on a price-supported or subsidized basis now and perhaps contribute to further development later (with adequate patent protection). In the plutonium business, this is all that could be expected, since there is only one prospective customer; but on the power-development side this reluctant attitude is significant, and does not seem to call very urgently for an immediate change in the law. This somewhat lukewarm response of those who have peered behind the wall of secrecy is perhaps the best indication available concerning the present promise of what is hidden behind it. It justifies serious attention to any other considerations that may be brought up before deciding whether to make a change in the legal status of atomic energy development right now.

Secrecy and Private Enterprise

The fact that four or five industrial groups have been privileged to look behind the wall of secrecy and decide whether they are interested illustrates how incompatible secrecy is with the kind of free competition that is the mainspring of our private enterprise system. This hiding of fundamental scientific and technical data is not at all the same thing as conventional industrial secrecy, which usually hides from a competitor only special technical developments and that only for a time, until patent coverage can be obtained. The atomic data hidden are much more fundamental because of the quite different situation. Not only is atomic energy an unusually advanced technical development, but the real competition is outside the jurisdiction of patent law—with the Soviet Union. If we stick to our present national atomic secrecy, we can only bring into the fold a very select team of carefully screened industrial groups, who can compete with one another inside the walls, but are not open to challenge by all comers. Such a restricted competition will not inspire ideas in obscure corners.

This is not private enterprise at its best. If we wish to use the lure of the effectiveness of the free enterprise system as justification for a change, we can do so consistently only if we can decide at the same time to go beyond the present stage of declassification and let down the bars of secrecy concerning the production of fissionable materials. This entails a further examination of the reasons for maintaining the secrecy, and of course it is possible to discuss this only incompletely outside of the wall of secrecy.

We may discern three main aims of the secrecy: (1) to hide details of the construction of atomic weapons themselves, (2) to avoid helping Soviet production of fissionable materials by informing them of details that they have not found out for themselves, and (3) to hide the size of our weapons stockpile. Information on bomb construction (1) is not affected by the present discussion of materials-production secrets. Those fundamental scientific data that might be affected and still bear on bomb performance are probably known anyway. Items (2) and (3) are closely related to each other, but it is probable that (3) is by now a more important motivation than (2). . . .

Size of Stockpile

While these questions should be decided on the basis of a more detailed and informed appraisal, it is clear that a desire not to reveal more accurately the size of our weapons stockpile is at least one important reason for keeping materials-production data secret. So it must be asked how much security would be lost if the size of our stockpile were to be revealed. The question was seriously raised some years ago by the late Senator MacMahon, who suggested that release of the figures would greatly help our planning. No very clear answer was publicly given to this suggestion, but no official release was made. There are obvious arguments on both sides. It is a usual military precaution to try to keep an enemy guessing about the extent of one's preparations. If the Soviet planners now know our total past atomic production only within a factor of two, and should be given data so they could know it within 10 percent, this might help them decide to divert more or less of their total industrial capabilities to atomic production, or it might help them decide when it is a propitious time for a surprise attack, if they were to do these things methodically rather than as forced by political crisis. It is possible also that it might dissuade them from attack or from planning for future attack. It seems very doubtful that this release of information would do as much harm to us by helping Soviet planning as help to us by encouraging public and industrial participation in our planning and production and more spontaneous development of our atomic science.

Secrecy-Limited Commercial Atomic Production vs. the Present Contract System

It appears then that we can approach the commercial participation in atomic production or power development, or both, either in the conservative spirit of keeping the secrecy situation as nearly as possible as it is at present and permitting commercialization on a limited basis, that might or might not cut costs, but would at least permit a few selected companies to get in on the ground floor of a speculative future industry, or we might lift the security lid and explore the possibilities of "free enterprise" in this strangely limited field. . . .

Current discussions seem to be concerned mainly with the conservative approach, a change in the law and in procedures to permit commercial participation, characterized by profit opportunities, but limited by the necessity of permitting only selected companies to operate behind the wall of secrecy. . . .

Long-Range Security Important in Present Decisions

It has been frequently emphasized that the growth of atomic stockpiles is making the technical aspects of the assurances essential to a disarmament agreement increasingly difficult as the years go by.[2] The problem is already so difficult that it is doubtful whether any solution will be technically possible after another decade or two, and it appears from the outside as though no branch of our government has yet taken the rather considerable amount of trouble to assess carefully what the technical possibilities of solution are and will be. It is to be hoped that this will soon be undertaken.

We as a nation seem to have lost interest in international agreements on atomic control. Largely because of the vehemence with which the Soviets turned down the one reasonable proposal we made at the outset of the atomic age, and the frustrating consistency with which they have reiterated their vituperative denial, most of us have concluded that there is no chance that they will under any circumstances change their minds, or that the Iron Curtain is so essential to the stability of their regime that it is idle folly to think of breaking through it by agreement. Of course there have also been other more concrete actions outside of the field of atomic negotiation to convince us of Soviet ill will toward the West. But in concluding that no change toward agreement is possible, we tend to forget that nothing is quite certain in the political future, that the intermittent vehemence of Soviet diplomatic manners is a matter of official habit with them, and that the Soviet leaders are realists enough to have made radical changes in policy when they found it expedient to do so.

The fact is that both they and we are faced with an impossible long-range prospect, unless we quickly come to agreement to limit or eliminate weapons of mass destruction. We sometimes hope that their economy or their method of political control will fall apart at the seams, and they hope and are even taught to believe the same of us. Once we both cease our daydreaming, we are faced with the prospect of an atomic arms race getting out of hand, with power of destruction too enormous to be brought under control even if the political climate should in some later day improve. The resulting unstable balance, the trigger-happy situation in which each side fears a surprise attack from the other and strives for ways to minimize the retribution for an attack, cannot reasonably be expected to last forever without the ultimate calamity. It is possible that our leaders will become convinced that agreement, something even short of an ideal agreement, would be preferable to that bleak prospect. And it is possible that the Soviet leaders, either independently or prompted by sufficiently wise proposals from our side, may come to see that agreement is the better course, perhaps with the solace that the contention may then be relegated to the ideological sphere.

The cool judgments of conventionally reliable planners are usually based solely on what is expected most probably to lie immediately ahead. We plan to jockey for position in the cold war, to strengthen ourselves in quest of short-range security as the most practical way to enter into the apparently endless long-range struggle. A sounder judgment would be made by giving some weight also to that fainter but immensely more desirable hope that agreement may be reached on some form of disarmament. This does not mean abandoning or seriously reducing conventional short-range security. With present tensions, that decision might too easily invite attack instead.

There are situations where a decision to foster the hope of agreement would probably impair short-range security only slightly or not at all, and it is here that the long-range problem should be carefully weighed in making the decisions. The question to what extent or how rapidly to replace conventional firepower with tactical atomic weapons permits conflicting opinions concerning the effect on short-range security, so this decision should perhaps be influenced by the realization that a rapid transfer to these weapons increases the technical difficulties of international control. . . .

These are some of the decisions in which the long-term considerations should be given some weight, and another is the decision discussed in this issue concerning commercialization of atomic production. Among the long-term considerations, the prospect of an agreed termination of the atomic armament race (as a part of a more general agreement) is not so remote as to deserve to be neglected.

Increased Industrial Participation and the Problem of Control

Liberalization of the conditions for industrial participation in the atomic production program could affect the prospects for making a transition to international control in two ways. One of the most crucial parts of the quest for international control is the problem of accountability of crucial atomic materials, and this problem might be either eased or intensified by the proposed change, depending on the nature of the change. Another crucial part is the problem there would be of "selling" atomic control to the public and its congressional representatives, even if a favorable agreement could be reached, and this is surely intensified by the establishment of further vested interests in the atomic explosive business. We have only to think of the political difficulty of removing farm price support, for example. . . .

Concerning the problem of materials accountability, it is important to keep the apparatus of production management in such a form as to facilitate submitting it to international inspection and ultimately international control. Back in the days of more unspoiled hope of easier international cooperation, this was one of the motivations in setting up the Atomic Energy Act in its present form. The revision of the act which is about to be proposed by the AEC for furthering "commercial development of atomic power" is said to be "predicated on the assumption that the UN is not likely to adopt international control of atomic energy, and therefore it is tailored to meet a purely domestic situation" (*New York Times,* April 8, 1953). Some press encomiums of this proposal stress only the possibility of commercial atomic power, overlooking the anticipation of industrial power as a mere by-product of plutonium sales to the government. Plutonium is produced at present by industry for the government on contract, and it is not clear that there will be any very great change in the type of control the government would have. With the present contractual arrangement, the government owns the plants and the crucial materials and industry operates them essentially on a cost-plus contract, which is not very profitable by industrial standards, if indeed profitable at all in anything but prestige. The industrial manager has an AEC manager, in another building, perhaps, but not very far from his elbow. Operations are carried out and records are orderly and available in a way not very different from the way it would be if the government could operate directly with its own employees. After the change, the company would operate for a profit depending on the company's ability to cut cost and probably on sustained government purchases. In either the conservative approach or with greatly reduced secrecy, the company would presumably own the plant. It it owns the fissionable material, this ownership would be nominal, limited so as to prevent conversion into bombs. If there is to be a change in control mechanisms, it would be in the direction of giving the company greater freedom of action. Perhaps the AEC would become merely a security watchdog, checking mainly on the inventory and quality of the plutonium, guarding against diversion.

The process of submitting to international control would probably be complicated considerably by the diversity created by this sort of arrangement.

It cannot be claimed that added complication would be very significant, compared to the enormous difficulties encountered at best. But more important than the greater complexity in the *process* of submitting would be the greater difficulties in the *spirit* and perhaps the legality of submitting to international verification of declarations and eventual control. It would be considered an invasion of private *rights* for the government to agree to such controls.[3] This is perhaps the most important respect in which the new proposal is "tailored to a purely domestic situation."

In either approach the establishment of vested interests would thus further dim the prospect of ending the arms race, but in the free enterprise with reduced secrecy approach this drawback would probably be more than compensated by the way the reduced secrecy would ameliorate the difficulty of materials accounting.

The most difficult technical part of a prospective international control plan is the problem of providing adequate guarantees to convince each side that the other is not hiding a secret stockpile, and this involves finding various clues to prove as accurately as possible how much crucial material the various atomic production plants must have produced. The simplest of these clues would be records, if one could trust them, and a free and open system not trying to hide the amount of production would lead to more dependable and interlocking records. Not only that, but we can convince the Russians more easily in the future by having let them collect clues as we go along, not requiring them to look only into the past. They may not yet be interested in real disarmament, but we may count on them to be interested in making this much preparation for it!

But why should we be interested in freeing this information now while the Russians are not? It is widely recognized that unilateral action as drastic as unilateral disarmament is futile as a quest for peace, but there is a strange sound basis for this particular unilateral freeing of production data, so far as it relates to the problem of instituting controls. This arises from the asymmetry of the East-West balance in conventional and atomic arms, respectively. It is generally agreed that we have a much larger atomic stockpile than do the Russians. In studying clues to past production, the atomic detectives will probably be left with something amounting to a percentage uncertainty in the size of the stockpiles. A certain percentage of our stockpile will be a lot larger, we assume, than about the same percentage of the Russian stockpile. So it will be the uncertainty in our stockpile that will be the chief technical deterrent to the writing of a satisfactory control plan. It is this uncertainty that would be reduced by our freeing of atomic information related to the size of our stockpile now, and as we go on producing.

Thus the difficulties of accountability would be made perhaps a little greater by the cautious approach to commercialization, but greatly reduced by reduction of secrecy in a transition to openly competitive private enterprise, and both types of change have the drawback of establishing vested interests. There is still the third possibility left—that the AEC should be encouraged to develop power under present contractual arrangements.

AEC Power Development

If the AEC should undertake development of power on an industrial scale, and then distribute it or sell it for distribution to industrial or urban users, there would again develop a vested interest in continuation of the power production, on the part of the users, but this vested interest would not be such a serious deterrent to a transition to international control. If the power were a by-product of plutonium production, some changes in plant would be necessary to convert it to a power-only plant, but the AEC would not have any profit incentive in plutonium production to tempt it to oppose the transition. And if "peace should break out," there would be no need for a large subsidy to keep the power plant going, because the terms of a disarmament agreement could reasonably provide that the government could use its surplus plutonium to run such plants under appropriate supervision (assuming this is technically possible), pending the development of economically self-sufficient power production. (This would also help dispose of the stockpiles.) Alternatively, if the purpose is to develop and perfect power-producing techniques by actually producing power, there is no need to distribute it, for the AEC is a large power consumer and could advantageously consume what it produces. Any technically possible economy in by-product power could be realized in this way, and this demand would terminate with the end of plutonium production. This also has the immediate political advantage that it does not compete with established power interests in the usual market—it merely partly deprives them of the expanding AEC operations as a future customer. . . .

We are thus faced with three ways to develop atomic power for industrial use: (1) development of a power-only plant or of by-product power by the AEC with the present type of contractual and legal arrangements, either with commercial distribution or preferably with the AEC itself as the power consumer;

(2) the conservative approach to commercial production, with little change in secrecy and very limited competition; or (3) making data concerning production generally available to all industry to encourage the open competition of the private enterprise system. On the basis of the limited information available, an overall judgment of short-range and long-range benefits to security would seem to suggest that we should choose one of the extremes and avoid the middle course. The middle course (2) seems to have no clear advantage over (1) in the short range, except perhaps as a lever in "practical politics," and has a clear disadvantage in the long range because of the vested interest it creates. The bolder course (3), besides introducing the tonic of open commercial competition, has the long-range advantage of greatly easing the problem of international fissionable material accountability.

NOTES

1. J. B. Beckerly, *Nucleonics* 2 (January 1953): 6.

2. D. F. Cavers, *Bulletin of the Atomic Scientists* 6 (January 1950): 13; 8 (March 1952): 84; D. R. Inglis and D. A. Flanders, *Bulletin of the Atomic Scientists* 7 (October 1951): 305.

3. The proposed Bricker amendment to alter and limit treaty-making procedures would tend to strengthen the legal status of such rights.

Nuclear Energy and the Malthusian Dilemma

[1971]

The Malthusian threat of a catastrophic collision between decreasing resources and increasing population has been used as reason for bypassing the critics and going full steam ahead with the present program of nuclear power development—most recently and thoughtfully by Alvin Weinberg, director of the National Laboratory at Oak Ridge (*Bulletin of the Atomic Scientists,* June 1970). The possibility of avoiding catastrophe is seen as a race between the population explosion—which will inevitably proceed further despite efforts to contain it—and the technological efforts to increase resources, of which development of new sources of power is a crucial part. If these are to proceed independently, there being some irrevocable number of generations before the population can be induced to limit its growth, then the crusade for increased power can be seen as a heroic attempt to buy time for the evolution of a stable population. It is then implied that no effort should be spared in the development of nuclear energy, even including the present proliferating generation of uranium-235 burners.

With much of this there can be no quarrel. Surely, avoiding both the extreme Malthusian catastrophe and the catastrophe of all-out nuclear war are deep concerns of us all. However, one need not and should not assume that the progress of the population explosion will proceed independently of the prospect of resources right up to some sudden day of judgment. If there is a strong coupling between them along the way—and I believe there is—this profoundly affects how one should develop and proclaim the hope of future resources in order best to avoid catastrophe.

Moreover, the ultimate Malthusian threat is not the only level at which there is concern for the environment. A population leveled off at a density leaving room for some of the amenities of extra space would surely be a happier one than a maximum population at the limit of survivability. If social and political pressures for population control can be intensified by premonitions of scarcity long before the point of catastrophe, the goal should be not only to avoid the catastrophe but to avoid it by a pleasant margin. If this be "tightening the Malthusian vice," as P. R. Ehrlich calls it, then enough of it to hurt only a little could be a useful supplement to other incentives for population control.

The drain on resources is augmented not only by the population explosion but also by the expectation explosion. We gluttonous and power-hungry Americans are leading the world in this respect and thus compounding the difficulties of the population explosion. The world looks to our standard of living as its goal for the future. Weinberg bases his projections for power needs on bringing the world up to our present standard of power consumption, but the aim of the present generation of U-235 burners is to help raise still further our standard of living and the world's goal. The explicit justification, by the utility companies that are advertising to increase power demand, is the need to help meet a demand for electric power that is doubling every ten years, or a per capita demand doubling every twelve years. In the perspective of Malthus this is the ultimate insanity, this assumption that the per capita electric power consumption of the most favored nation on earth must double each decade or so as part of an economy that knows no god but an ever-increasing GNP.

Even professional economists in this country, with few exceptions, analyze the economy in these glowing terms. Were it not for the advent of nuclear power, it seems likely that the cost of reducing visible pollution and of reclaiming lower grades of fossil fuel would by now have started to make a dent on their thinking and that of the body politic. But they are being deluded into continued dreaming by the as yet unjustified promise of pollutionless and perpetually plentiful nuclear energy.

The drive for the expectation explosion comes from each industry wanting to do its thing at a profit. Every baby born, like every new gadget that can be sold, and each lap in the arms race is a potential boost to the market as long as there is no limitation on resources. Money talks, and relaxed abortion laws, tax disincentives to suppress excessive consumption, or high-priority approaches to arms limitations are hard to come by in the political arena. A mild power shortage as a reminder of the Malthusian problem of the future could exert a restraining influence and might even inspire suitable political or economic intervention.

If and when dependable and safe nuclear energy sources based on really plentiful fuel supplies become available, they will of course contribute importantly to the possibility of supporting increased numbers of human beings in some degree of comfort and productivity. Research toward this end is being carried out and is urgently needed. It is deserving of sharply increased financial support. Nuclear energy presents the possibility, but not the promise, of abundant power in the future. Reasonably trouble-free fission breeder reactors and fusion thermonuclear devices are exciting possibilities but they are chickens that must not be counted before they are hatched. As Weinberg has explained, the former are the nearer prospect, while the latter, if ever feasible, would be the more satisfactory solution to the world's energy needs because of their greater fuel supply and less serious radiation problems.

R & D Costs

The American public is paying, and will pay rather generously, for nuclear energy research and development (R & D). But the way the money is apportioned between research and development is determined by an abhorrence of socialism that is hardly in keeping with Malthusian concerns and is not necessarily permanent in our already partially socialized economy. Most of the research and initial development of new concepts is paid for through taxes financing national laboratories and their subcontractors. Most of the development and construction of nuclear power plants is paid for by the public through electric bills.

The initial passing of the torch from the Joint Committee on Atomic Energy (JCAE) and Atomic Energy Commission (AEC) to big electrical concerns and utilities in the mid-fifties was motivated by the thoughts that the R & D on uranium burners was far enough along, that industry has the capacity to get big things done when the remaining R & D is not too speculative, and that nuclear electric power thus promoted would brighten the name of the atom somewhat tarnished by military applications. The utility companies had the incentive of getting in early and preventing the government from producing nuclear electric power so as to preserve this as a domain of private enterprise. They even refused to let government nuclear pilot plants feed electric power into commercial distribution grids. Yet, before it would take the leap, industry required substantial arm twisting by the government, such as the subsidy represented by the Price-Anderson Act to avoid liability for disastrous accident. Optimistic reactor builders helped lure electric utility companies into the nuclear business by supplying equipment initially below cost, as it turned out, with the hope of making up some very considerable losses at higher prices later on. When the conditions were made attractive enough for one utility to enter the field and install a reactor, then they were attractive enough for many to do so. The result is unnecessary duplication of big experimental reactors.

The goal is power priced competitively with that from fossil fuels. If this were attainable straightaway, the duplicated experimentation would, in principle, have no direct monetary cost to the public. It has not been attained even in the best of cases. The most modern nuclear plants have not yet caught up to the most modern fossil fuel–fired plants in low mills per kilowatt hour. Being truly experimental, most reactors have been plagued with costly troubles, sometimes delaying successful operation for years, and some have so far been expensive white elephants. For this the public does pay heavily in higher rates on electric bills. Power billing rates are pegged to production costs so that the monopolistic utility companies will not make more than a fair return on their investment. This regulation of rates means that the way to increase profits is to increase power consumption, regardless of expensive mistakes that may be made in choice of power sources and regardless of what is done to the environment or to future fuel supplies. It also means that temporary losses sustained by utility companies and reactor manufacturers, resulting from early gambles on optimistic projections, will eventually be passed on to the public as the actual cost of new reactors is reflected in readjusted electric rates.

The actual cost to the public may be hard to assess, for there will also be rate increases due to dwindling fossil fuel supplies and the necessity of a shift of emphasis—from oil and gas to coal that is increasingly expensive to mine as mining conditions improve. Particularly because of this shift it seems most desirable that the resources being expended on duplication of nuclear burner and breeder experiments should be redirected to R & D in efficient production and use of fossil fuels with minimized environmental pollution.

Unfortunately, from the long-range Malthusian point of view, the experiment in nuclear power production thus being supported by the public is the wrong experiment. It is experimentation to develop the reliability of water-moderated uranium-burning reactors, whereas the hope for plentiful nuclear power lies in breeder reactors, if it is to be in fission reactors at all. Apart from temporarily catering to our craze for power, reactors of this new generation, technically refined though they appear, are too primitive to be worth developing for themselves. They utilize essentially only the power from the U-235, seven-tenths of one percent of the total uranium metal that should be useful with a successful breeder, and they will use it up at such a rate that the easily obtainable supply will last for only a decade or two.

In Weinberg's opinion "the experience we shall be gaining from the present generation of reactors—in waste disposal, in handling and control of radioactivity—will be most important" for the introduction of breeder reactors, say within twenty years. This part of the experience I claim we can gain better and more safely without the present generation of reactors but by spending some money directly for this experience instead, where it needs to be spent.

Economy is paramount in the commercial production of nuclear power in competition with fossil fuel and any merely incidental experience in improving waste handling comes slowly indeed. The method of reducing high-level wastes to a glassy solid was developed so long ago that proof of its economy and practicality was announced from the Chalk River Laboratory at a Geneva conference in 1958. Yet its use on more than a small demonstration scale is a thing of the future, one to which the Atomic Energy Commission seems finally committed with the purchase of a salt mine in Kansas for storage. This method of disposal is said to cost only 0.04 mills per kilowatt hour of electricity produced. It is surely money that should be spent, and presumably will be, to put an end to the tank farms for more than temporary storage. But in the meantime we have almost 200 million gallons of high-level liquid waste to give us experience with safe disposal, and that should be enough without the waste from the new generation of reactors. If part of their purpose is to produce waste, let's not have them.

As for the handling and control of radioactivity, the pressure on cost in the present generation of reactors is leading to malpractices that should not be condoned. For example, it is making us take unnecessary chances with the diversion of bomb-grade nuclear material. The 1967 act making enriched uranium private property has led to appalling practices in this respect.

While most reactors use uranium enriched by only about 3 percent in U-235, some achieve an average enrichment of about this amount by using some fuel rods enriched to about 90 percent interspersed with a larger number of unenriched rods. Thus 90 percent enriched uranium, potential bomb material, has become a regular commercial item. It is now routinely shipped by private companies in commercial trucks and aircraft with hardly more caution than if it were so much iron. These matters are not publicized, but it has been said by an authority that large amounts of about 90 percent material, enough for several bombs, have gone astray and been lost for as much as a week. While our diplomats and the technologists backing them up are valiantly trying to devise means of avoiding the proliferation of dangerous nuclear materials in the spirit of the Nonproliferation Treaty (NPT), the recent commercialization of the handling of nuclear materials seems to be an experiment in how long nuclear bomb material can be exposed to the wiles of gangsters without catastrophe. This is experience we should not be having for the sake of economy in water-moderated reactors, and do not need for breeder reactors. There are more technical aspects of handling and control for which routine experience is doubtless rather useful, but not experience in the parsimonious economy that is encouraged by building and financing too many reactors in competition with coal. We should ultimately have nuclear power when it is so much needed that we will be ready to pay for caution. In the meantime we should be paying directly for research and pilot-scale development of safe and efficient methods of processing and handling nuclear materials, not indirectly for the kind of large-scale commercial development in which economy demands cutting corners.

Another aspect of experience, emphasized by another Oak Ridge scientist, Walter Jordan (*Physics Today*, May 1970) is perhaps a little more relevant, but not useful enough to motivate running large numbers of reactors of a dead-end generation. This is experience in avoiding a catastrophic runaway accident, or explosion. Assessing the risk of such an event is very undependable. Weinberg says: "Surely the chance that such an event will ever happen is very small. Yet one cannot prove negative propositions of this sort." He points to past experience as a guide, but past experience is very limited indeed compared with expected future experience and even so is not all sweetness and light. Actual catastrophe has been avoided, but how close have we come?

For the chance of such an event Jordan tentatively names the figure one ten-thousandth per big reactor per year, which would mean a likely one catastrophe (that he hopes would be mild) per century with a hundred reactors operating. For what it may be worth, this is a far cry from Weinberg's "chance that it will **ever** happen is very small." But one ten-thousandth is little more than a number drawn from a hat and in another man's opinion one thousandth might seem more realistic, expecting one bad accident per decade of hundred-reactor operation. Present experience is only enough to make the figure one hundredth seem very doubtful, insofar as it can be transferred at all to larger reactors making somewhat greater demands on materials.

False Sense of Security

If we go ahead with the proposed hundred or more big water-moderated reactors in the next decade or so and run them for a decade without serious mishap, we will have made the figure one thousandth doubtful and Jordan's one ten-thousandth plausible only for water-moderated reactors. Some of this experience may be transferable to whatever type of breeder reactor may by then have been developed, but not all of it. Long experience with the integrity of pressure vessels under conditions of intensive radiation damage may be relevant. More important, perhaps, may be the assessment of the likelihood of human error where unusual care is needed in a commercial environment. Yet the demands for perfection in a trickier breeder reactor of the future might be so much greater that this experience would do no more than give a false sense of security.

Thus our proposed huge splurge in water-moderated nuclear power plants in the next decade has rather little to do with the development of the future types of breeder reactors that will make a substantial contribution to alleviating the Malthusian threat. The present power plant program—both burner and breeder—is even counterproductive in the sense that the extra money and effort that will be spent on proliferation in numbers would be much better spent on encouragement of innovation in research and development of new types of breeders and on intensifying fusion research, as well as research on cleaning up fossil fuel techniques on which we will in any case be primarily dependent for decades.

But then comes the question of being practical in the light of political and economic realities. How does one effect the transfer of funds from one program to the other? Those engaged in promoting long-range nuclear power to alleviate the Malthusian dilemma might argue, as Weinberg seems to imply, that we scientists and engineers have no control over how the nation spends its money. Money to cover mistakes in building and managing a lot of water-moderated reactors is available through electric power bills that the public has to pay without making a decision, whereas tax money to support a Malthusian-oriented program mostly in national laboratories requires voting by the political representatives of the economy-minded public. The practical anti-Malthusian may thus feel it prudent to applaud and participate in the present trend to overbuild, no matter how undesirable its side effects or misleading its slogans, for the sake of the crumbs that fall from the orgy to the anti-Malthusian effort.

Review Overdue

But it would seem far more prudent to recognize that a review is overdue of the decision to let nuclear development pass out of the national laboratories, where it was progressing well, and into commercial channels with their tendency to unbridled Parkensonian proliferation, a decision made in an early period of euphoria in the early fifties, when the seriousness of many difficulties and side effects was not fully appreciated. It was made before there was even our present rudimentary information on biological concentration of the low-level radioactive wastes that flow downstream from reactors and fuel-processing plants into the sea. It was made when it was still thought that there was no cancer caused by the bursts at Hiroshima and Nagasaki, before knowledge of Sr-90 in milk, before there were even our present inadequate hints of the extent of infant mortality and genetic damage from radiation and when one could still postulate a threshold below which low levels of radiation might be harmless. It was made before the accidental meltdown in the first commercial breeder reactor, operating at one-tenth power in its first year, gave cause for worry about whether disastrous accidents of similarly completely unexpected origin might occur, in the many much larger reactors of the near future. It was made before realistic assessment could be made of how fast we would go through easily accessible uranium supplies with burner reactors. And finally, from the Malthusian point of view, it was probably made without consideration of the harm that could be done by postponing through rosy hopes the onset of a gradual adjustment of our profligate economy to the ultimate necessity of leveling off.

Rather than to support an industry that prematurely advertises nuclear power as "clean, dependable, economic, and safe," it would seem more prudent to try to convince the appropriate congressional leaders to reorient the program toward making the slogans true through increased emphasis on future-oriented research and development instead of encouraging, subsidizing, and licensing further present deployment. The clock cannot be turned back, but it should be possible to modify trends.

Power-mad America

Much of the very rapidly increasing demand for electric power is artificial, induced directly by over $20 million worth of advertising per year by the power companies and indirectly by countless millions more spent by other companies to advertise a host of products of electric power, such as soft drink and beer cans. Freedom to advertise lies at the heart of our free enterprise system, though in the case of products of dubious safety such as cigarettes it is regulated. Freedom of the utility companies to advertise power consumption is the result of a well-financed campaign to keep in the private domain a monopolistic type of public service industry that in most other industrialized countries is a government enterprise.

Of course the advertising would not pay if the people did not like the products being sold. They lead to greater convenience, greater luxury; but in many cases it is convenience that we can ill afford both in the light of prospective near-term rise of cost of all types of fuel and in the light of the long-term Malthusian threat and the example we set to aggravate it.

Long-term Cost

A prime example is the much-advertised use of electric power in place of fossil fuel–burning furnaces to heat homes. The first cost is low, for heating ducts and furnaces are replaced by comparatively cheap resistor units in floor, walls, or ceiling. It is attractive especially to speculative home builders as a way to increase profits. Most new houses on the market in some northern parts of the country are built this way. But the long-term cost of such heating is high because it is an inherently inefficient way to use fuel. A scientific principle first formulated long ago by the French physicist Carnot teaches us that there is an inevitable loss in converting heat into electric power to be converted back into heat, as compared with using the fuel to produce the useful heat directly. In the most modern oil-burning electric generating plants a thermal efficiency of 40 percent has been achieved by use of very high temperature boilers. (The best nuclear plants are considerably less efficient essentially because neutron-irradiated materials cannot stand such high temperatures.) This figure means that every unit of heat that heats the all-electric home is matched by at least one and a half units of heat that heats the river or lake or bay on which the generating plant is located. This means that more than twice as much fuel is burned to heat the home as would be consumed if it had an oil furnace. This means roughly twice as much atmospheric pollution, even making some allowance for a possible difference in the burning efficiency. In addition, it means that every house that is built to be heated with electric resistors in the walls reserves for the lifetime of the house the right to pollute thermally a body of water with more heat than heats the house.

In the far future when we may be almost entirely dependent on nuclear fuel for power, electric heating of homes may be necessary in climates where solar heating is insufficient. But now only about 1 percent of electric power is nuclear and even the projected splurge of U-235 burners would take this figure to only about 20 percent by 1980. For the next decade or two the choice is essentially between fossil fuel burned in the electric plant for inefficient home heating, and fossil fuel burned directly in the home. Under these circumstances it seems to be excessive worship of private enterprise to permit installation of electric home heating by means of cheap resistor units. Electric heating by means of a heat pump—air conditioning in reverse—is another matter, but its initial cost is high, particularly in cold climates.

A Costly Habit

Electricity is a wonderful servant, but one that should know its place. Its convenience in reasonable amounts of illumination and in running most of the many electric motors and electronic devices in the home has become almost indispensable in our comfortable way of life. The amounts of electric power thus used are considerably less than in heating (or in some cases air conditioning) a home and need not be discouraged. Air conditioning, on the other hand, particularly because of the peaked nature of its demand for power, does have undesirable ecological consequences and its excessive use should be discouraged.

Another serious example is our excessive use of metal cans, whose production demands large amounts of electric power. Something like half of the demand for electric power is for metallurgy, and much metal is wasted for rather frivolous purposes. With high-grade iron ore becoming scarce, larger amounts of power are used, particularly in substituting aluminum for iron. Much aluminum is used in the canning of beer and soft drinks. There was a time when most Americans reached for a glass of water from a tap or occasionally went down to the drug store for a soft drink. The new generation—environment buffs among them—seems to have the habit of reaching for a can of cola instead. This habit, like the all-electric house, is a costly and unnecessary commitment for the future use of power.

One need not advocate that Americans decrease their comforts to save electricity nor even that they stop increasing their per capita demand for electricity. One should perhaps demand that those Americans near the top of the economic heap no longer increase their individual power consumption. Some increase in per capita consumption is well justified to spread comfort and convenience to our less fortunate citizens, of whom we still have too many. But that is not what is causing the doubling every ten years. On a worldwide scale, getting at least one electric bulb burning in every primitive hut to help keep people up until they are ready to sleep is a worthy goal as an element of population control. But again, this is not what our proposed reactor splurge would be accomplishing.

Prudence Requires

The goal of the proposed dubious and dead-end program of water-moderated reactors is to contribute 20 percent to our swollen electric power production which, a decade from now, is expected to be double today's per capita production. If, through wise regulations limiting advertising and certain types of consumption, and by readjusting power rates, we should keep the increase in the next decade down to 60 percent instead of the expected 100 percent, we could dispense with nuclear power completely and still have no greater demands on fossil fuel production of electricity. We should be able to limit the demand more than that and still have a healthily, though less rapidly, expanding economy.

Prudence, then, would seem to require efforts to induce power-mad America to exercise a little restraint and cease to lead the revolution of rising expectations so rapidly out of sight for the world's teeming billions.

We should avoid pell-mell expansion of a premature nuclear industry while sparing no effort on R & D to advance the practicability and safety of the future nuclear power sources that may be a large part of the answer to the Malthusian threat.

Energy Gluttony and Overkill

[1972]

The basic assumption behind energy planning in the United States is that the per capita consumption of electricity will go on doubling each decade. The population increase, while alarming, proceeds only about one-tenth as fast and accounts for only a small part of the 100 percent increase in electric power every nine or ten years. At this rate we are expected to use about eight times as much electricity in the year 2000 as we did in 1970.

According to projections made by the Atomic Energy Commission, about half that enormous production of electricity will be derived from nuclear energy, the other half coming from fossil fuels—coal, oil, and gas, but mostly coal as oil and gas supplies dwindle. Even with the projected assist from nuclear energy there would be about four times as much electric power produced by fossil fuels at the end of the century as now.

Both nuclear and fossil fuel production of electricity have serious drawbacks. Both pollute the environment today, and, by depleting the natural resources of our wonderful planet Earth, both preempt for our present convenience materials that will be sorely needed for necessities of life in the future. The nuclear method also has some special counts against it: century-long threats of radioactive pollution and the making of nuclear materials that might go astray.

Cheap and plentiful electricity has long been a goal of both government policy and industrial pressure. Electricity, it has been said, should be as plentiful as the air we breathe, and as cheap. In the early days of nuclear power development there were dreams of power so cheap that it would not pay to install switches to turn off lights.

Optimistic hopes like this played a part in our original commitment to the pursuit of industrial nuclear power. These hopes have not been fulfilled, but the cost of electric power from coal, oil, and gas has, until recently, been going down while the cost of other commodities went up. This has been achieved partly by engineering advances, but largely by tolerating environmental costs that do not appear on an electric bill. These include a rampage of devastating strip mining for coal and the raw untreated pollution from the smoke stacks of power plants.

Nuclear power was born as a federal government baby, nurtured in national laboratories, and served up to industry on a silver platter. The techniques of burning the fossil fuels, coal and later oil and gas, grew up with the industries that they made possible, nurtured sporadically by many companies and with no central source of direction or support for research and development (R & D). Consequently, the money and effort usefully devoted to improving the technology of mining and burning coal have been pitifully small compared with the resources devoted to reducing the hazards of nuclear power generation. Both coal (the only fossil fuel that will be available over a long period) and nuclear sources are pollutants. But the coal enterprise is the more starved for money to clean up its production and burning by known methods, as well as for R & D to improve those methods.

Nuclear fuel now accounts for less than 2 percent of our electric power. It is expected that a decade from now, when electric power will have doubled, nuclear sources will account for about 20 percent of it. Of the 100 percent energy increase in that time, 60 percent will be from fossil fuels and 40 percent will be nuclear. If we should have a moratorium on new nuclear plants now and let the fossil fuel development march on as expected, there would be a 60 percent increase in electric power rather than a doubling in the next decade.

This would still be a rapid expansion, about six times as fast as the population increase. All that our plunge into the uncertainties of nuclear power is doing is to keep up the rate of increase at 100 percent rather than letting it relax to 60 percent in the next decade. One might think that a 60 percent increase would be more than enough, even for our power-ravenous economy with its worship of a burgeoning gross national product.

Giving up nuclear power—stopping the construction of new plants completely—would not mean giving up some of the fine electric appliances in the home to which we have become accustomed. Primarily it would mean less conversion of industries to production methods gulping huge quantities of electricity.

Whether or not we give up nuclear power for the next decade, it would be desirable to clean up the burning of fossil fuels to reduce environmental contamination. Cleanup devices will increase the cost of electricity. The demand that has led to the rapid doubling of energy output has been partly artificial, spurred by advertising by the utility companies which claim they need to expand quickly to meet the demand they create. This inflation of demand should be stopped—one company has already stopped it—but an increase in utility rates for the sake of the environment would also help to pare down demand to fit a less rapidly expanding supply. An inversion of the rate schedule, making the first hundred kilowatt-hours per month the cheapest and higher levels of demand more expensive, would also help to suppress extravagance and would avoid putting an undue burden on the small consumer.

The net effect of the thousand or so reactors of both types in the next thirty years—if they materialize—will be to prolong and increase the disparity represented by the oft-quoted fact that we Americans, about 6 percent of the world's people, consume about 35 or 40 percent of its goods and power. On a worldwide basis, these reactors will make the rich richer and more extravagant while doing precious little for the poor.

The rapidly increasing drain on resources comes not only from the population explosion but also from the "expectation explosion." We extravagant and energy-hungry Americans are leading the world in the explosion of rising expectations and thereby compounding the difficulties of the population explosion. The world looks to our standard of living as its goal. The aim of our new nuclear reactors, and the doubling of electric power each decade, is to help raise our standard of living still further and thereby the world's goal.

There are simply not enough natural resources, no matter how much power is used, for the world to attain the U.S. standard of living. We should leave something with which the underdeveloped can improve their lot. A continually rising economy is riding for a fall in power output. A nuclear moratorium would be an effective way to begin to decrease the rate of rise in power output, and would avoid the serious difficulties we are likely to encounter in the premature expansion of an insufficiently developed technology.

One of the most important of these difficulties is that we will be producing large quantities of highly radioactive waste products when we have not yet developed a satisfactory means for keeping them out of the environment, where they will persist for centuries. Burying them in salt mines has been proposed but not proved out.

Another difficulty is the unassessable likelihood—probably small for any one nuclear power plant—of a radioactive catastrophe from a runaway accident. If the economy becomes completely dependent on the power from many nuclear plants, and such an accident should occur in one of them, killing many people in a nearby city, it would be difficult to decide whether or not the other plants should continue to operate. A further difficulty is the routine release of small amounts of radioactivity from power plants and particularly from nuclear fuel–processing plants. Accidents in the transportation of radioactive spent fuel are also a danger.

But perhaps foremost among the dangers is the possibility that the nuclear materials produced in the power plants might be diverted to the clandestine making of atomic bombs. All these nuclear power plants make plutonium. A single uranium-235–burning power plant of the type now being constructed—not even a breeder—creates about a thousand pounds of plutonium a year, enough to make more than a hundred atomic bombs. With hundreds of such plants in operation, and plutonium worth four times its weight in gold, a dangerous black market supplied by thefts is bound to develop. That will mean that the stuff to make atom bombs will be available to almost anybody, any petty dictator or underworld chief. The potentialities for criminal mischief that could trigger general nuclear war are ominous indeed.

Even now, before we have reached a largely plutonium-based economy, there is potential trouble. Some of the water-moderated reactors use, along with other fuel rods, a few fuel rods of uranium sufficiently highly enriched to make atomic bombs. There is already commercial traffic in this material. In the interests of cheap electric power, the regulations for care in shipping the uranium rods are no more stringent than for a registered letter. It has been demonstrated that the rods can be hijacked easily in transit, and there have been several examples of shipments of significant amounts going astray. Fortunately, so far, they have been eventually recovered. For the same stuff in military hands, where security rather than economy is the word, the regulations are more stringent and the possibility of diversion somewhat less.

The International Atomic Agency, abetted by our own AEC, is making an effort to control this situation but it is not sufficient to cope with the threat that is developing with so many new reactors. Just as we have gone ahead with the arms race before giving high priority to arms control, just as we have entered into the production of great amounts or radioactive fission products that must be safely sequestered for centuries before developing adequate means to do this, so we are embarking on producing, for commercial use, enormous amounts of the stuff that atom bombs are made of before having instituted substantial controls to prevent its diversion to dangerous and illicit uses.

A primary reason that the nuclear power plant program seems to be riding like a steamroller over the objections of the environmentalists is not only that people with nuclear skills and interests in and out of the government like to do nuclear things, not only that there is money to be made in building nuclear plants and selling their power, but also because scores of industries can increase their profits by using cheap electricity.

Both in nuclear weaponry and in nuclear power generation we are forging ahead but at great—and unnecessary—risk. In nuclear weaponry, we and the world would be safer if we called a halt without going on to more sophisticated arms systems, and safer still if we had stopped at a lower level, or could work back to a lower level of missiles on both sides deterring one another, for we have already piled overkill on overkill.

Now that each of the nuclear giants seems to understand well the importance of not starting a nuclear war, the danger that large-scale nuclear war might break out seems to stem at least as much from the possibility of trouble starting with nuclear weapons in the hands of irresponsible small powers or cliques as from the possibility of direct political confrontation between the nuclear giants. There is something grossly incongruous about the way we have used the excuse of staying ahead of the Soviets with our unnecessarily superior overkill to justify huge expenditures on the profitable weapons business while we have, at the same time, let the economic lure of nuclear power generation dull our sense of caution about producing the materials for a nuclear weapons black market.

We overkill-mad and power-ravenous Americans are leading the world a merry race, but a dangerous one. When it comes to both the military atom and the peaceful atom, we should learn the value of restraint.

Nuclear Power: Rasmussen Reviewed

[1976]

The 1974 official Reactor Safety Study WASH-1400 (the Rasmussen Report) uses analytical methods capable of assessing the relative magnitudes of various hazards, and these lead to some useful recommendations in the effort to make reactors safe. But the report goes further and gives some highly questionable estimates of the absolute magnitudes of hazards. That the main purpose of doing so was a desire to assure the public of the safety of reactors seems clear from the uses to which the report's conclusions have been put. In the publicity and in the executive summary which introduces the report, the very small estimates of the risks of major nuclear calamities are accepted at face value and are compared favorably with many other public hazards, as though comparison with increasing numbers of other risks would make the small estimates more credible. In neither place nor in the testimony which succeeded in getting Congress to renew the Price-Anderson Act (providing government insurance and limited liability in the event of a nuclear accident) is there mention of the serious warnings and doubts that have been introduced into its appendices by conscientious participants in the study. Nor is there adequate mention of the limited data base for the hazard estimates, which excludes those "unscheduled events" in the commercial reactor experience which have come closest to calamity.

Of all the warnings in the appendices, perhaps the most damning is one sequestered back on page XI 3-61. It cannot be interpreted otherwise than to say that, essentially, the study did not take into account the possibility that serious accidents might result from the multiple failure of components caused by deterioration with age (more than five years), since all the results of the study apply for only five years. One wonders how a study that claims validity for only five years—let alone one neglecting such a cause of failure in an apparatus as subject to deterioration as a nuclear reactor—can be used as reassurance of reactor safety through the eighties.

In order for it to be believed that such a surprising statement is made in the report, the statement must be read in context. Criticisms and comments were solicited for the WASH-1400 draft report so that these might be anticipated and answered in the final report. The surprising statement is the last sentence of a response which addressed two related comments at once, the entire context being as follows:

"COMMENT 3.2.10: Examples were given of actual incidents that involved several sequential human or equipment failures. The comment questioned the ability of the study to predict such events using the methodology employed in WASH-1400. (*Union of Concerned Scientists; The National Intervenors*)

"RESPONSE: In performing its assessment, the study reviewed not only the examples cited in the comment but also many other sources of pertinent data. The study's analyses were not meant to be taken out of context and extrapolated to different situations or different sequences. Sequential failures must be treated by sequential methods: alternatively, it is necessary to identify, by the use of methodology similar to that discussed in sections 3.1.2.1 and 3.1.2.2c of this appendix and in Appendix 1, single based causes that govern the sequences of failures. In one instance cited, aging was used as an example of a common mode of failure. It should be recognized that the study did not include extreme aging considerations since the applicability of its results is limited to only the next five years."

While responses to some of the comments have led to improvements, the critical reviews have found many serious flaws which were not corrected in the final report. Many of the responses are inadequate, and the report seems to have come off the worse for wear in its attempt to meet the critics. Indeed, a most frightening commentary on nuclear safety is the thought that all the talent behind a two-million-dollar study could produce nothing more genuinely reassuring than WASH-1400.

The gratifying record of no known deaths to the public from the operation of commercial power reactors in this country is often quoted by utility companies as proof of the safety of nuclear power. With the invalidity of WASH-1400 admitted even in its own appendices, this statistic is about the only valid indication of reactor safety we have. Gratifying as the record is, it means only that the likelihood of calamitous accident is in order of magnitude no more than one in one hundred or at best one in one thousand per reactor-year.

This is nowhere nearly good enough as a basis for continuation of the nuclear energy program and particularly its massive taxpayer support. It is reason instead to throw similar heavy government subsidy into the promotion of energy conservation and much more rapid development and deployment of renewable alternative energy sources.

Power from the Ocean Winds

[1978]

The prospect is looking up this year for having the wind generate important amounts of electric power. For thirty-six years Vermont held the record for the world's biggest windmill: one-and-a-quarter megawatts fed into the commercial grid. Late in 1977 this record was shattered in Denmark, where a cooperative school group with meager finances built a modern two-megawatt wind turbine with three blades 177 feet in diameter. That record will not stand long. Two of these huge machines are being built in this country, with completion expected about the end of the year. One, which will produce two megawatts of power, is being built in North Carolina under the federal wind power development program; the other, which will produce three megawatts, is being developed independently in California.

On a more modest scale, the federal program has two 200-kilowatt windmills operating and a third scheduled for early 1979. There is a 140-kilowatt windmill in Washington state and a 200-kilowatt one in Massachusetts, both built by private firms. A government-sponsored 200-kilowatt, vertical-axis machine is currently operating in Canada. Several private firms are now fine-tuning earlier successful prototype models for mass production in smaller sizes—under 50 kilowatts—and the prospects are bright for rapid expansion in this field.

But there is one vast opportunity for reaping power from the wind that is receiving practically no promotion at all. That is offshore wind power.

Offshore Wind Power

The power of the wind reaching a windmill is proportional to the third power of the wind speed: thus, doubling the speed of the wind means an eightfold increase in the amount of power produced. For this reason it pays to go where the strong winds are. In the West, this means the western part of the Great Plains and certain favorable wide mountain passes through which the wind pours. On land in New England the power of the wind averages only about 150 watts per square meter, whereas in the first hundred miles or so off the Atlantic Coast, above the moderate depths of the continental shelf, the figure ranges from 400 to 700 watts. There is space out there to moor tens of thousands of floating, megawatt-scale windmills. That is where it should pay to put them, if this can be shown to be feasible. But in this respect Yankee ingenuity and enterprise have failed us: no one has been willing to try.

A test demonstration of offshore wind power was seriously proposed in 1972 by William E. Heronemus, professor of engineering at the University of Massachusetts. In the following years the officials of the federal wind energy program repeatedly rejected the proposal and they have continued to oppose the idea ever since. After preliminary design studies, Heronemus proposed testing a full-size unit such as might be used in large numbers off the Eastern seaboard.[1] His six-megawatt design consisted of three two-megawatt wind turbines on a spar-buoy type flotation unit. The main hull was to be a submerged hollow sphere with a ballast sphere below it. It would carry a tower above it to support the turbines aloft. Only the hollow legs of the tower would normally be awash in the waves and the whole unit could lean over, yielding to severe storm winds. The ambitious nature of this proposal is clear once it is realized that at that time no wind turbine as large as two megawatts had yet been built on land.

Heronemus also provided a glimpse into the future in proposing that the power from thousands of these units could be converted to hydrogen at sea. Part of the hydrogen would be put aside for windless periods in expandable storage tanks under the pressure of the deep sea and it would all be conducted ashore by pipeline, either to be used directly or to be converted into electric power in fuel cells. Such an arrangement should ultimately be cheaper than a simpler submarine cable to shore.

The Federal Program

On land, too, the federal energy program has been lamentably slow to reap the bounty of large-scale wind power, especially in view of the great national need for diversification of power sources. In building the first big windmill in Vermont during 1939–41, a small industrial firm achieved a megawatt-scale machine in less than two years from the time of its first interest in wind power. The federal program has taken six years, from 1972 to 1978, to do the same thing, while achieving a 100-kilowatt-scale windmill in four years and fostering many smaller projects and paper studies along the way.

In its "research and development" program on 100-kilowatt and larger windmills, the federal effort is doing nothing really new. It is merely designing and building, with important engineering improvements, windmills of the same general type as the one built in Vermont years ago, with a single two-bladed rotor on a fixed tower. This is the conservative approach, the sort of approach one should perhaps expect in a low-priority government program, particularly one sponsored by an agency heavily committed to other energy sources. A less conservative program with higher priority could have been deploying hundreds of big windmills in extensive wind farms by now. In this perpsective, the failure to explore offshore wind power can be seen as merely one aspect of a general conservatism.

In the first few years of the federal program, as money began to be available that might have been used for an offshore test, the program director, Louis Divone, gave as his reason for opposing it his belief that a failure in so innovative a project might discredit wind power in general, and the modestly supported federal program in particular. Therefore, he held, it would be better to wait until there had been many successful demonstrations on land to establish confidence.

Some congressional leaders have been impatient with the pace and scope of the wind power program. Appropriations have repeatedly exceeded budget requests. The fiscal 1976 energy budget included an amendment on offshore wind power introduced by Senator Kennedy. The legislation, passed in summer 1975, required the authorities to initiate a feasibility study and then to submit to Congress in a future budget request a proposal for a megawatt-scale offshore wind power demonstration project.

The response to this requirement has been dilatory. Such studies are carried out by contract with appropriate groups. The choice of contractors could quite conceivably influence any conclusions about feasibility. An industrial giant with competing vested interests might yield to a bias toward an unfavorable result, while an independent group of able engineers might be more apt to exercise originality in finding ways to make the proposal feasible. Perhaps two or more parallel studies could have been made by somewhat different groups to explore the possibilities. As it was, the sole offshore feasibility contract (for almost a quarter of a million dollars) was awarded to the firm most heavily involved in nuclear energy, Westinghouse.

Selection of the contractor and awarding of the contract took more than a year—for a study that was supposed to take about a year and be finished in the fall of 1977. Half a year later the results were known within the Department of Energy but after another half-year are still not available to outsiders. Thus it will perhaps be too late to influence the shaping of the fiscal 1980 budget, which will be considered by Congress four years after the congressional mandate.

Economics of Wind Power

Last May 3, Sun Day, President Carter announced his sudden decision to give solar energy development an extra $100 million. Wind energy's share was $20 million. Despite this unexpected increase, a policy review led to a decision within the Department of Energy that "offshore is out." The reason given was that the new study showed that offshore wind power, while perhaps feasible, does not appear to be economically competitive with other power sources. The general argument is that land-based wind power, according to Department of Energy projections, promises to be barely competitive, if that, leaving no room for the rather high extra costs of offshore mounting.

The present rejection of offshore wind power as an important energy resource for the near future thus stands squarely on the Department of Energy's economic perceptions, not on lack of technical feasibility. There is compelling reason to discount the DOE's evaluation of wind power economics. The department seems to have been strongly influenced by dealing with large space-oriented organizations stooping to mundane, if still challenging, land-based wind power development. The federal program's first medium-sized windmill, a 100-kilowatt machine carefully built by NASA in Ohio but with an inexcusable design error, was planned for test purposes. Its inflated price of about $10,000 per kilowatt therefore should not be considered typical.

Subsequent big windmills in the federal program are running about $2,000 per kilowatt, still a high figure. By way of contrast, a small engineering firm, Charles Schachle and Sons of Moses Lake and Seattle, Washington, is building a three-megawatt windmill (the first machine of its size) in a windy region for Southern California Edison, at a contract price of $350 per kilowatt, including development costs and profits. The 140-kilowatt prototype, designed for scaling up to the larger size, has been running successfully since its completion over a year ago.

Another private company, U.S. Windpower of Burlington, Massachusetts, plans to market a forty-kilowatt windmill, weighing two thousand pounds, for an initial price of $15,000; the company expects to reduce the price to perhaps half that once the windmill is in quantity production. The starting price is $375 per kilowatt. The machine is a refinement of the federally sponsored twenty-five-kilowatt windmill that has been heating a house in Amherst, Massachusetts, for several years. Once these windmills are produced in quantity, their price will be close to the price, as well as the weight, of a good small car.

Detroit produces the equivalent of more than ten million small cars a year. That large a year's production of these windmills would have a rated generating capacity of 400 gigawatts (400 million kilowatts), about the rating of 400 big, one-gigawatt steam power plants. With a factor of 40 percent to allow for the unsteadiness of the wind, the average power generated would be about 160 gigawatts or the equivalent of about 270 large nuclear power plants operating at a typical average 60 percent of full power. This power could be produced from only one year's output of an industry the size of the auto industry or, perhaps more practically, from ten years' production of an industry a tenth that large. That is a lot of generating capacity—and a measure of what we could expect from the wind if we made this a matter of national priority. It would also create a lot of jobs. With energy storage[2] to provide steady power and with transmission lines from windy regions to big cities, these low windmill costs could be expected to make wind power cheaper than power from future nuclear plants.

The Task Ahead

These practical commercial costs are important for the prospects of offshore wind power in two ways. First, they are so much lower than the costs the DOE has been experiencing for its windmills in somewhat gentler winds that the DOE's economic reason for rejecting offshore wind power has to be regarded as unsound. However, the judgment about such potentially important research and development should not be made on purely economic grounds in any case. Secondly, those forty-kilowatt windmills will probably soon be in production in at least limited quantities. This should make it possible very shortly to construct a floating offshore demonstration unit using a dozen or more of these tested turbines practically "off the shelf."

Though harvesting the strong offshore winds involves the extra cost and problems of flotation and mooring, it does not require the rather expensive yaw mechanism that keeps a land-based windmill facing into the wind. A floating unit would automatically face into the wind by riding on its mooring the way an anchored ship does. Incidentally, it is for this reason more practical at sea than on land to construct a unit carrying not just one large wind turbine but rather many smaller ones that will benefit sooner from the economies of mass production. They may also be better able to withstand the buffeting of a storm.

The thought of a large array of floating windmills out over the horizon where the strong winds blow is an exciting prospect. It could supply a large part of the electric power needed by the urban East—a completely clean energy source with no fuel consumption and no cooling-water problems.[3] It requires a lot of doing, but first of all it must be shown that it can be done.

The task is not an easy one. It poses formidable engineering challenges but they should not be beyond the capabilities of modern engineering. Besides the vibration problems common to all windmills, there is the gyroscopic problem of a big rotor on a swaying base. The rotor would be stopped in a big storm but the demands on the structure and mooring for surviving a storm are severe. The worst storms out on the Grand Banks are fierce. Waves over one hundred feet high have been recorded at the Texas Tower off Nantucket. There is even some difference of opinion as to whether a well-designed wind power unit could survive the storm and icing conditions there. This very doubt adds to the importance of having at least one floating windmill in the ocean as soon as possible, testing and, one hopes, demonstrating the feasibility of offshore wind power. It is a question that can only be finally settled by experience, not by design studies on paper.

Such an investment involves some risk of expending effort and money on research and development without having it pay off. However, such venture enterprise is needed in all fields; without it there can be little real progress. Government subsidies and tax advantages have so distorted competition in the energe field that venture capital is not attracted to new undertakings. This suggests that venture capital for wind power must come from government sources if this technology is to expand rapidly. It is important to carry out the advanced research and development now, establishing the offshore option without waiting for rising fuel prices to remove all doubt about the ultimate economics.

The long-awaited report on the offshore feasibility study should be available soon. Being a new and extensive study of the problem, it should contain new insights. It is possible that its conclusions may be largely negative. They should not be accepted without critical reappraisal. Offshore wind power presents too great an opportunity to be dismissed for less than compelling reasons. We should not forego a critical test because the design problem looks difficult, or on the basis of premature economic judgments. As a nation we have the technical capability to do great things. We need to diversify our power sources and to stop using so much fossil fuel. We have the industrial capacity to bring wind power on line within a decade as an important source of energy. The offshore test is a small but important part of the needed effort.

NOTES

1. W. E. Heronemus, "Pollution Free Energy from Offshore Winds," reprint, Eighth Annual Conference and Exposition, Marine Technology Station, Washington, 1972; also "Wind Power: A Near-Term Partial Solution to the Energy Crisis," in *Perspectives on Energy*, ed. L. C. Ruedisili and M. W. Firebaugh (New York: Oxford Press, 1975), p. 375.

2. The most practical, benign, and economic storage currently available is pumped hydroelectric storage with one or both reservoirs underground.

3. It could also serve as a regional wind furnace, the supply from strong winter winds being well matched to cold-weather demand for heating.

Epilogue: Then and Now

This collection of papers is but a very small part of the great volume of thought and writing by many people engaged in the contention between two opposing views of the arms race. In one view, a bit over-simplified, the Soviet Union is so bent on aggression that it would risk a very severe degree of nuclear damage to itself if it thought it could do even worse damage to the West, and we need an enormous degree of overkill and even ability to conduct nuclear war to suppress such a venture. In the opposing view, represented here, the Soviets are much less aggressively inclined than that, being primarily concerned with avoiding being invaded or attacked again and with developing their creaking economy aided by what they can extract from their satellite nations, and only secondarily with spreading their outworn ideology while competing with the West for commercial advantage in the third world by means short of risking nuclear war. Whatever may be their intent, a much lower level of nuclear weaponry is seen as less dangerous and ample to deter aggression. Adherents of both views aimed all along to avoid nuclear war but differed sharply on how best to avoid it, whether by continu-ing the arms race indefinitely and even being prepared to fight a protracted one or by promoting mutual restraint. Within this contention the specific issues have come and gone and these articles address some of them, one after another, while the broad outline of the contention remains the same, as indeed do some of the issues as they return to haunt us again.

The Ultimate Threat

But now the nuclear threat is seen to be even more menacing for the future of humanity than was appreciated earlier. Then the threat of nuclear weapons was recognized as a threat of unspeakably terrible destruction and human suffering but not to the extent of annihilating our species. Ever since the grisly demonstrations at Hiroshima and Nagasaki the knowledge of the effects of blast, direct radiation, firestorm, and lingering radioactivity have provided reason enough to give our highest priority to the avoidance of the terrible suffering of large parts of the present human generation in nuclear war, not only through these direct effects of nuclear weapons but also through the subsequent disruption, starvation, and disease. The threat of genetic damage to future generations has added to the concern. Now, new insights reveal that all-out nuclear war could so damage the global environment as to exterminate at least the higher forms of life on earth, including mankind. In contemplating the significance of the possible obliteration of large segments of life on earth, it is impressive to recall how living things have developed in ways that could not be repeated: how life and its global environment have evolved together, each contributing in essential ways to the evolution of the other, while there developed also a high degree of interdependence among living species. Some three billion years ago, a quarter of the way back to the earth's beginning, the atmosphere consisted of water vapor and unpleasant gases such as methane and ammonia, with no free oxygen, and in this early environment the complex molecules essential to the beginning of single-celled life could be formed spontaneously. Without oxygen, there was no ozone layer to filter out the sun's ultraviolet rays and these contributed to the process. Early microscopic life, shielded from ultraviolet rays within the sea, produced oxygen for the atmosphere. This made possible the evolution of higher forms of life, both by providing air to breathe and by making an ozone layer to reduce the otherwise deadly ultraviolet radiation to tolerable levels. Life and the environment have since continued to interact in more detailed ways, the varied forms of life evolving in response to changes in environment very gradually over millions of years.

Man has existed only a brief million years or so of all that time, slowly improving in skills and thinking. In the last ten thousand years he has been learning agriculture and ways of living together in large groups, ways that changed little until recently in their dependence on human muscle, animals, and wind in sail for power. Great strides were made in art and philosophy but the earth was seen as most of the universe until four centuries ago when the solar system came to be understood, and then the grandeur of the galaxy. But only in this century has the true enormity of the universe been recognized. Early in the century science also peered into the realm of the very small to recognize protons and electrons as constituents of atoms and molecules. Since then there has been a deluge of discovery far surpassing any of the past. As recently as a half-century ago the span of knowledge extended only from galaxies to protons and electrons. Now it reaches much further in the direction both of the very large and of the very small, from quasars to gluons, and there have been fantastic advances in the atomic, molecular, and biological realm in between.

Just as we now arrive at this sudden flowering of knowledge that is a crowning achievement of the long evolutionary process, the process is threatened with extinction. For in this same dramatic half-century—a mere moment in the long span of history—man has acquired the means to destroy himself and there is no assurance that this will not happen.

The newly appreciated threat to the environment posed by nuclear war includes two separate threats, a super-sunburn threat and a deep freeze threat, either of which could do us in. The super-sunburn threat arises from newly discovered modes of sensitivity of biological species to ultraviolet radiation. These introduce the alarming possibility that all-out nuclear war with present levels of weaponry might not only damage future generations but eliminate them completely by inducing catastrophic breakdown of the complicated symbiosis of global ecology. This has been brought to public attention eloquently by Jonathan Schell in the first part of his book *The Fate of the Earth* (New York: Knopf, 1982).

While life and the environment have been evolving together very gradually over millions of years, many interdependent species have thus adapted to current levels of the ultraviolet rays in sunshine. Nitric oxide resulting from nuclear explosions can, to an unknown extent, deplete the protective ozone in the stratosphere and in all-out nuclear war it could suddenly increase the ultraviolet radiation, leaving the interdependent species weakened with no time for adaptation. Among the many organisms involved are birds and insects that might be blinded, and thereby doomed, plants whose photosynthesis would be impaired, and microorganisms at the base of the food chain in the sea. With many of its parts weakened, the whole interdependent system might collapse. It cannot be known that such catastrophic breakdown would occur. There is so much uncertainty in the magnitude of the individual effects and the mechanism of their synergistic interdependence that it cannot be known that it would not occur. Further research may help narrow the uncertainty. If nuclear war ever starts, it cannot be known that it will not go all-out. The first bomb so detonated, even the existence of the stockpiles, put life on earth, or at least the higher forms of it, seriously at risk.

This nitric oxide threat, with all its uncertainties, is a threat of too much radiation, in this case ultraviolet radiation somewhat similar to X rays, and this is a long-term threat. The other threat, the deep-freeze threat of a "nuclear winter," is a shorter-term and more certain threat of too little radiation penetrating the atmosphere, this time the visible and infrared radiation that keeps us warm. This would result from the smoke of many simultaneous fires in burning cities and forests that could be much more effective than were the dust and fumes that caused a "year without a summer" from Krakatoa. It would cool the earth to subfreezing temperatures, probably exterminating many species and perhaps all higher forms of life. The threat from smoke and the threat from nitric oxide together would constitute a global one-two punch to life.

The horror of nuclear war as a threat to our own lives and way of living is reason enough for doing whatever we can to try to avoid it. In the course of everyday life, no additional incentive is needed. The new dimension of concern for the continuation of life on earth may seem remote and even unduly sentimental or sensational. Yet for all who have a philosophical or religious appreciation of the true wonder of all that we see about us, the trust that we hold in our hands for assuring the continuity of life into future eons should be seen as a deeply sacred trust, still more important than assuring the well-being of just our generation and those of our children and grandchildren. Short-term survival is of course essential to long-term survival, but the long-term aspect is the more important from a global evolution point of view. Throughout history, the responsibility and function of each generation of each species has been to provide the next generation. Now our responsibility is for the ages. Though to some it may seem remote from practical affairs, this view should influence attitudes toward current arms control negotiations, providing added reason to make compromises for the sake of progress toward long-term goals.

After all the progress in human values that has been made, it seems utterly tragic that mankind, rather than being content with pursuing further development in friendly cooperation, is divided into three main groups, two of them facing each other angrily with ever-increasing nuclear arsenals while competing for influence in the third. The confrontation becomes a complicated one with many detailed nuances but we may see several general influences contributing to its intensity. To the hard-liners among our policymakers, no further explanation is needed than the recalcitrance of the other side, and less polarized explanations may be seen as "soft on communism." Yet, despite the great differences between the United States and the USSR in personal freedoms, standard of living, and individual participation in government functions, there is a great deal of similarity in the way the two sides contribute to the arms race. The policymakers and wider circles on each side include two opposing factions, hard-liners and soft-liners or hawks and doves. Each side has a large and influential military establishment steeped in the tradition of fighting wars and a supporting industrial complex with a vested interest in further arms production. On each side the hawks pick up statements made by leaders of the other side suggesting aggressive intent and emphasize them to promote support for further armament. The two military establishments are self-perpetuating but need each other—and even exaggerated perceptions of each other—to prosper.

Two important influences propelling the arms race are competition and the technological imperative. Competition is a spur to great efforts in many pursuits. Competition between primitive tribes, then city-states, then between nations has led these groups to devote great efforts to the martial arts and to valor in battle. Competition is the driving force in athletic contests, often with a spirit of rivalry between institutions whipped up for the occasion. In scholarship and the arts and sciences, pursuit of pure excellence to achieve a goal is a strong force, but in most work it is strengthened by competition for prestige. National leaders throughout history have sometimes felt the need for an enemy or a potential enemy to keep a nation alert and unified. And so it is with the current arms race. It takes two to conduct an arms race, and the potential enemy must be portrayed in lurid terms to make him worthy of the great competitive effort perceived to be required. Even among the subunits of the military structure on each side there is competition. Among our armed services, among the three legs of the strategic triad, the competition may be healthy as a spur to effort and performance, but is unhealthy as a spur to the arms race. The pressures on Congress seem to require that a new weapons system for one must be matched by budgetary concessions to the others.

Science and technology seem to know no limits. In pure science, the goal is knowledge for its own sake, with faith that unforeseeable future applications will in the long run be beneficial to mankind, that understanding of the workings of nature will facilitate adaptation to nature, and this is as it should be. But in practical technology, where immediate goals can be seen and evaluated, and particularly in military technology, a similar perception unfortunately seems to prevail. There is a feeling that the advance of technology cannot be stopped, or that it would somehow be immoral to try to stop it, regardless of the consequences a particular development may have. This technological imperative paces the arms race. While military desires do shape the allocation of funds in the application of current technology, strategic doctrine is commonly modified to accommodate the application of each successful new advance.

Developments thus motivated make the nuclear confrontation in the eighties differ substantially from that in earlier decades when most of the articles in this volume were written. The development of strategic missiles with multiple independent warheads or "reentry vehicles" (MIRVs) has in principle, though perhaps not in practice, spelled the end of the invulnerability of the land-based leg of the strategic triad. As accuracy becomes sufficient, one missile is presumably able to knock out more than one missile silo, even with allowance for some failures. On the receiving end, having several warheads on the missile in the silo means that one attacking warhead is able to knock out several warheads. Our introduction of MIRVs thus increases both our vulnerability and that of the Soviets, destabilizing the confrontation. Just before we achieved MIRVs, the Soviets proposed agreeing to ban them but our procurement momentum prevailed. This is one of many opportunities there have been for arms control that we have missed and now, with MIRVs aimed at us, our military planners feel the cost of our shortsightedness.

In earlier decades the accepted strategic doctrine was deterrence through mutual assured destruction, called MAD though it was and, at some level of deployment, remains the most reasonable option in a world cursed with the nuclear threat. However, it has its problems of credibility. A national leader must convince the adversary that he would probably retaliate if attacked even though, just after a limited attack, it may be in the national interest not to escalate the carnage. The technology of increased missile accuracy makes possible a response more limited than deliberate city destruction, the so-called surgical response against purely military installations, though collateral population damage in most cases would still be catastrophic. Strategic doctrine on our side follows the lead of technology and includes this as a way to make response more credible, but with the illogical option of "decapitation," or destruction of Soviet leadership and command structure on which it depends for exercising restraint. Efforts are even being expended to develop our capacity to fight a protracted nuclear war. All this makes sense, in the hawkish view, only on the basis of the slim hope that calm and reason will keep passions in check and avoid attacks on cities even in the midst of a nuclear exchange. The immediate aim is what is known as "escalation control," based on the notion that the two sides can agree to stop a nuclear war after a selective exchange of a few relatively "small" nuclear blows in spite of having been unable to avoid starting one. This seems unlikely particularly since the Soviets will have none of such doctrine, but instead appear to see nuclear war as almost inevitably total. Yet our policy to pursue the current buildup of new weapons systems, rather than effectively negotiating restraint, is based largely on that slim hope.

It is claimed by its proponents that a counterforce strategy is more moral than a countercity one. It is difficult to assess morality in the intrinsically immoral domain of nuclear war. In terms of the direct effect of the first blow, it is surely less immoral to threaten an accurate "surgical" strike on a remote military target, where collateral damage should be moderate, than to threaten a nuclear strike on a city, the very thought of which is most repugnant. Yet morality of a nuclear strategy should rather be judged by the influence it has on the probability that unrestrained nuclear war will occur. The probabilities are imponderable, a matter of judgment. If, for example, it should be estimated that the chance is more than fifty-fifty, or one-half, that an initial counterforce blow would escalate into a city-destroying war and also that an initial attack is twice as likely to occur if both sides are equipped to pursue their goals with counterforce threats than if deterrence rests on a rough balance of mutual assured destruction, then an eventual attack on cities would be more likely if both sides have counterforce capabilities than if they do not. While these simple numbers are only illustrative, in my judgment the conclusion holds that the existence of specialized counterforce capabilities increases the odds that cities will be destroyed. The lowered threshhold with a counterforce strategy makes the threat more credible, but less to be feared, and more likely to be carried out. I thus find mutual assured destruction the more effective deterrent, and as such the least immoral policy available. Furthermore, morality is on the side of improving the chance of avoiding nuclear war by stopping the arms race now and soon reversing it, seeking reduced tensions and greater stability at a lower level. While net reductions could be made during limited counterforce deployment, most counterforce advocates give no more than lip service to serious efforts to end the arms race.

The cruise missile represents an especially irrational pursuit of the technological imperative, for its massive deployment, if carried to completion, spells the end of any possibility of ending the arms race by verifiable arms control, leaving nuclear war as a likely alternative. This small but potent missile is more innovative than the MX and more detrimental to long-term security. The MX, to be deployed in the United States, and the Pershing 2, being deployed in Europe in response to the Soviet SS-20, are merely important modernizations, in terms of greater accuracy, of the huge ballistic missiles on which deterrence has been based for the last quarter-century. The modern cruise missile, though related to World War II buzz bombs, is an entirely new and more subtle nuclear delivery system. The heightened danger to the stability of the deterrent stems from its smaller size, making verifiable arms control almost impossible to achieve in the future and making it more easily available to third world countries with limited resources despite its advanced electronics.

Until the advent of the cruise missile, indeed until the Reagan reelection in 1984 practically assured its massive deployment, I nursed my personal hope that the decades-long efforts to end the arms race would not only end the upward spiral but start a downward trend toward a relatively safe minimum deterrent level and eventually perhaps even on to a nuclearly disarmed world. Now, with the prospect of unlimited deployment of modern cruise missiles, there is less basis left for such hope. The most daring stance that the prevailing attitude in Washington permits doves there to take is for strictly verifiable arms control leading to substantial reductions. Massive deployment of cruise missiles makes that objective obsolete. A much greater change of attitude will be required to make progress in the presence of cruise missiles than without them. It is more difficult to imagine a scenario for civilization's evolution through a century or so of nuclear confrontation without nuclear war in a transition to an era lasting many millenia not cursed by the threat of nuclear war. Without cruise missiles, my hope scenario would be to achieve a verifiable minimum deterrent at a low level in less than a century, nuclear war being avoided by decreasing tensions accompanying reductions along the way, then development of an ABM shield and subsequent reduction to disarmed deterrence for the millenia. With cruise missiles a minimum deterrent might still be attained, but one in which it is more difficult to develop confidence. It would again consist of perhaps a hundred verifiable, single-warhead ballistic missiles, but now amidst a sea of an unknown number of slow cruise missiles whose perceived military function would be only to follow through and complete destruction that would already have reached unacceptable proportions. Thus the swift ballistic missiles could be depended on to deter an attack by either kind of missile and the exact numbers of cruise missiles stashed away should not much matter. If a defense against cruise missiles could be developed and afforded it would be helpful, but not until it is close to perfect should a balanced introduction of ABMs be considered. The final transition to an ideal world without nuclear weapons would not be possible unless a way could be found to eliminate cruise missiles. Failing that, it is much more reasonable to expect a minimum deterrent situation to last on into future millenia than to expect a perpetual arms race to avoid eventual catastrophe. The distant goal of achieving the long-term stability of minimum deterrence thus remains a strong incentive for taking initial steps of arms reduction now.

Throughout the decades covered by this volume there have been occasions when a new weapon or weapons system became developed to the deployment stage and the Pentagon has concentrated its promotional efforts on it alone, implying that without it we are practically defenseless but that with it we would be safe. Thus opponents of the continued arms race have had to resist one proposal after another. If one is knocked down, another crops up as the arms race continues. The clean bomb, the neutron bomb, the missile gap, and ABMs have each taken their turn in the limelight. More recently, the emphasis has been on the modern cruise and Pershing 2 missiles and their role in maintaining a separate nuclear balance between the United States and the USSR in Europe, independent of the global balance.

While the European theater has long been crucial to the determination of nuclear policy, particularly the development of tactical nuclear weapons and the doctrine of first use in case of invasion with conventional weapons, this insistence on a separate balance gives it special prominence. With their SS-20s the Soviets in the early eighties have very substantially increased and modernized their intermediate-range force targeted on Europe. It was previously inferior both to the French and British forces, most of them submarine-based, and to the American submarine-based missiles deployed off the coast and functioning as part of the NATO defense. NATO had reasonably decided long ago that it was more effective to base them in submarines than on land, making them invulnerable to swift intermediate-range attack. With the availability of the more accurate cruise and Pershing 2 missiles for land deployment, that decision and even the existence of the missiles at sea was forgotten in administration rhetoric that implied we would be lost without the new missiles. When the number of Soviet SS-20s in the European sector reached three hundred, the sales pitch claimed that the score in Europe was three hundred to nothing, a gross misrepresentation of the situation. The new land deployment can be seen to have the advantage of making the deterrent more credible by permitting the United States to participate in a European nuclear war in the optimistic hope that it would not escalate to involve the American homeland. It is, however, destabilizing, reducing the flight time of the first American missiles arriving in a counterforce strike on the USSR from half an hour to seven minutes, thus encouraging the Soviets to be ready for instant launch on warning.

A more alarming obsession is the projection of the arms race into space. Advances in space technology have brought benefits for our everyday lives as well as fantastic dreams. The marvel of stationary-orbit satellites provides such convenient long-distance communication that many of our terrestrial activities have become dependent on them. Surveillance satellites have both civilian and military uses. They will be vulnerable to attack as killer satellites are developed, yet we seem unable even to try to avoid this development by mutual agreement with the Soviets. They have suggested such an agreement and we are again the ones who insist on pressing on into the dangerous unknown. But the expenditures in the Buck Rogers domain go further, to include the hope of destroying all ballistic missiles of a massive attack on launch by laser beams from space, for example. Viewed realistically, this is an unattainable goal, requiring unlikely perfection of intricate systems and fantastic efforts to boost the required fuel into space. Even if all this were technically feasible, the space stations on which it would depend would be vulnerable to destruction as the first step of a ballistic missile surprise attack, and as if that were not enough, there are other means besides ballistic missiles for delivering nuclear weapons. Yet Congress has gone along with the Pentagon's fascination with such projects. At appropriations time in Washington and during the election campaign the president has even had the effrontery to bemuse the American people and confuse world statesmanship by painting the hope, as though it were real, that such means will rid the world of the scourge of nuclear weapons if we spend enough money on them as part of the arms race.

Plans for large ABM systems were abandoned and such systems abolished by treaty two decades ago largely because the general conclusion was accepted that a defense against nuclear weapons would have to be at least 99 percent perfect to prevent unprecedented damage, and such perfection of a complicated system is just not in the cards, particularly of one that cannot be previously tested on a full scale. The new technology has not changed that basic fact. The discussions in chapter 4 still apply.

Of the Pentagon's promotional sallies, two have notably failed: the shelters for civil defense (second paper of chapter 6) and ABMs (chapter 4). Each failed largely from its prospective ineffectiveness but not without crucial help from active opponents of the arms race. A related failure of a promotion by the Department of Energy was Project Plowshare, or nuclear explosive engineering, that could have accelerated weapons proliferation. Despite the inadequacies demonstrated earlier, shelter programs as well as ABM programs are being promoted again in Washington.

These papers have been intended to influence government officials both directly, hoping that some read, and through informed and aroused public opinion, as have the writings and activities of many other scientists and other individuals and organizations similarly motivated. Some of these scientists have also had special influence through having served at length in important government posts. While there have been successes in special issues, the effort has failed to achieve its broad aims, for the arms race goes on. Relentless pressure from the hard-liners in the Pentagon and their supporters has by and large prevailed. The very fact that the negotiations did continue in parallel with the arms deployments must be due at least in part to that aroused public opinion.

Back when the early papers of this collection were written, many of us who already appreciated the power of nuclear weapons were especially deeply alarmed. I felt the chances were perhaps less than even that I would escape nuclear war and live to old age. Now that we have survived another third of a century, for which I feel very grateful, we are inclined to take this as grounds for optimism. That the world has muddled through this long really proves nothing about the future, but fosters hope. A pessimistic view is that we have built a house of cards higher and higher, one that is increasingly likely to collapse. In a more heartening view our experience is said to show that deterrence works. This is like concluding after the Yanks have had a winning streak that they always win ball games. Actually, either the experience shows that deterrence has worked so far or deterrence was unnecessary because no major power has incentive enough to want to wage war, conventional or nuclear, against another major power. I have now come to believe that the latter is almost certainly the case, that the Soviets have no sufficient incentive to attack, but that under present circumstances it is nevertheless prudent to retain an adequate deterrent as a hedge against misjudgment, one strong enough to squelch any grandiose ideas of easy conquest yet modest enough that it neither exacerbates relations nor unduly drains resources needed for the smooth functioning of society. If it is strong enough for this aspect of deterrence, it would also be strong enough to prevent the Soviets from gaining influence in the world through credible nuclear threats.

In the first of these articles, ''A Deal before Midnight?'', I expressed a belief in Soviet aggressiveness that is still widely held when I wrote, ''It [the atomic bomb] served us well in the short period of our monopoly. It is probably responsible for the freedom of Western Europe from Soviet domination today [1951].'' In retrospect I can see no likelihood that the Soviets would have tried to overrun western Europe with their massive military manpower and surfeit of tanks. These they had built up after great losses during a period of enormous devastation and tragic suffering to repel the relentless German invasion for which they had been unprepared. Their problem was to rebuild their country and consolidate gains that included the division of Germany as insurance against recurrence of such tragedy. It is not plausible that they would have started a new war after having finally finished that very costly one if there had been no atomic bomb.

Now that both sides do have nuclear weapons, I believe the same is still true, that the Soviets do not have any incentive, commensurate with the costs, to start a war against a major power or any undertaking that would risk nuclear war. Not only would nuclear retaliation cost them untold suffering, but they also have a stake in the prosperity of the West. Economic interdependence is even more in their self-interest than in ours. They know that their system could not produce yields in the plains of Kansas to feed them as well as ours can. They have enough trouble preserving their political domination over their buffer-zone satellites without trying to extend it to the world.

We cannot be sure how the Soviets might behave if they had a monopoly on nuclear weapons as we had in the late forties, so that they could make demanding threats without risking nuclear retaliation. We need a nuclear deterrent only as insurance against their taking advantage of such a situation. Now in the nuclear age mutual deterrence, with arsenals powerful enough that each side can inflict unacceptable damage on the other and with as small a "nuclear club" as possible, is the most stable situation that can be achieved, short of some more remote ideal, to preserve international relationships more or less as they are.

As for how powerful is powerful enough, anyone who truly appreciates the scale of nuclear destruction, thinking vividly of the Hiroshima experience from a bomb about a hundred times less powerful than a typical H-bomb, should conclude that if each side had the assured capability of delivering one H-bomb to the largest city of the other, or surely if it were ten bombs on ten major cities, this would be quite adequate to deter, whatever the incentive. Even if we should choose to retain the option of "surgical" strikes to avoid initial direct threats to cities, this would require accurate missiles but relatively few of them. Such a low level of deterrence is not to be considered a practical goal until attitudes change a great deal, but it is a measure of how unreasonable it is to dicker over detailed balance of large numbers of weapons and thereby avoid indefinitely ending the arms race and achieving substantial reductions.

But even if there might develop a strong urge to strike in a crisis, a hundred deliverable warheads should be plenty to deter and more would make comparatively little difference in the dissuasive effectiveness of prospective damage. With this in mind, the main reason to let the numbers on each side remain over a hundred would be to retain the perceived special status of the two superpowers, if China, France, and Britain would not agree to proportionate reductions.

In the early postwar years when I did not appreciate that lack of any incentive worth the risk of nuclear war, Stalin was in charge and there was fear that the Soviets under his brutal leadership might attack even though their land was in shambles. The passions of war lingered on and there was a general feeling, which I shared, that the quarter-century between world wars I and II represented a natural period for the recurrence of major war.

Memories were then still fresh of how World War II started with sudden surprise attacks by both Germany and Japan. We should appreciate that each expected a quick victory at relatively small cost in the prenuclear era, Hitler with his "blitzkreig" or lightning-war end run past the Maginot Line, and the Japanese General Staff by eliminating the U.S. Pacific Fleet at one blow and thus dominating the Pacific while the attention of the United States was distracted by war elsewhere. Each led a militarily and economically strong, but geographically small, country feeling itself unjustly confined by history to too small a living space, seeking "lebensraum" and a "place in the sun."

The outlook of the Kremlin is very different from that of Hitler's Nazi leadership. What they have in common is quite different ideologies antithetical to ours. Nazism grew out of Germany's smarting under its defeat in World War I, involving loss of territory to France. Its military objectives were expansionist to redress past losses. Soviet military attitudes are generally recognized to be very different, primarily defensive in reaction to vivid memories of the 1941 Nazi invasion at the start of World War II and a history of earlier invasions. Whereas Germany in the thirties was seeking revenge for defeat in the previous war, the Soviet Union enjoys the satisfaction of having redressed its grievance with Germany in the victory we helped it attain, splitting that country in two while establishing a band of buffer states as a protective belt around its borders. Its problem is not living space but adequately developing the vast space it has. Its adventures in remote parts of the third world pose a serious problem for us, perhaps more economic than military, but they have at least stopped short of risking nuclear war. In that regard nuclear deterrence may have worked indirectly as a moderating influence. Any such influence as there may be would not depend on the current high degree of overkill capability. Deterrence at a much lower level would do as well.

The perception of the Soviet Union formed in those early postwar years persists in some quarters today, having been periodically nurtured to spur enthusiasm for the arms race. Throughout most of the intervening years the

Soviets have been somewhat reluctantly following our lead in each new lap of the race while suggesting calling a halt. They, more than we, seem to value the relative safety of stable deterrence. If we had accepted some of their proposals in the past, such as avoiding multiple warhead missiles, we would be in less danger now. If we were to go along with some of their present proposals, such as avoiding a new arms race in space, we could avoid added dangers for the future. If we can ever free ourselves from our bondage to the profit motive in weapons development, there will still be problems of detail to be thrashed out in the task of retiring most nuclear arms and reducing the likelihood of the final catastrophe, but they all should become manageable if the two sides could view and treat each other without exaggerated fears of evil intent. We have a planet to inhabit together, we hope for a very long time, and much that needs to be done.

Supplemental Papers
[Other publications concerning public affairs by David Inglis]

Articles in Journals

"Tactical Atomic Weapons and the Problem of Ultimate Control." *Bulletin of the Atomic Scientists* 8 (1952): 79.

"The New Working Paper and U.S. Atomic Diplomacy." *Bulletin of the Atomic Scientists* 8 (1952): 132.

"U.N. Disarmament Commission Proceedings." *Bulletin of the Atomic Scientists* 8 (1952): 271.

"H-Bomb and Disarmament Prospects." *Bulletin of the Atomic Scientists* 10 (1954): 41.

"Armament Decision in a Democracy." *Bulletin of the Atomic Scientists* 12 (1956): 196.

"Prospects for Stopping Nuclear Tests." *Bulletin of the Atomic Scientists* 13 (1957): 19.

"Clean and Dirty Bombs." *New Republic* 136 (June 10, 1957): 10.

"Future Radiation Dosage from Weapons Tests." *Science* 127 (1958): 1222.

"Allaying Suspicion of Test Ban Controls." *Bulletin of the Atomic Scientists* 15 (1959): 425.

"A Test Ban Is Still Possible." *New Republic* 142 (March 7, 1960): 11.

"Excessive Fear of Test Ban Evasion." *Bulletin of the Atomic Scientists* 16 (1960): 168.

"Congressional Hearings on Technical Aspects of Test Control." *Bulletin of the Atomic Scientists* 16 (1960): 105.

"Testing and Taming of Nuclear Weapons." Pamphlet No. 303. New York: Public Affairs Pamphlets, 1960.

"On Resuming Nuclear Tests." *New Republic* 146 (February 12, 1962): 18.

"Test Ban Default? Obstacles to a Crucial Agreement." *Council for Correspondence Letter* (1962): 12.

"Lessons from the Resumption of Nuclear Testing." *War/Peace Report*, January, 1962, p. 12.

"Disarmament After Cuba." *Bulletin of the Atomic Scientists* 19 (1963): 18.

" 'No Cities War' or Disarmament?" *Bulletin of the Atomic Scientists* 19 (1963): 17.

"The Test Ban: Too Little?" *Correspondent* 29 (November–December, 1963): 13.

"Of Arms and Man." *Correspondent* 30 (January–February, 1964): 27.

"Nuclear Weapons Control on the Way to Disarmament." *Forensic Quarterly* 38 (1964): 307.

"Chinese Bombshell." *Bulletin of the Atomic Scientists* 21 (1965): 19.

"The Region-by-Region System of Inspection and Disarmament." *Journal of Conflict Resolution* (1965): 187.

"Missile Defense, Nuclear Spread, and Vietnam." *Bulletin of the Atomic Scientists* 23 (1967): 49.

With C. L. Sandler. "Special Report on Plowshare." *Bulletin of the Atomic Scientists* 23 (1967): 46.

"Nuclear Pollution and the Arms Race." *Progressive,* April, 1970, p. 17.

With Allan R. Hoffman. "Radiation and Infants." *Bulletin of the Atomic Scientists* 28 (1972): 45.

"Why the Arms Race?" *War/Peace Report,* July/August, 1973, p. 10. Also in the *Congressional Record,* December 20, 1973, p. 23469.

"Wind Power Now." *Bulletin of the Atomic Scientists* 32 (1975): 21. Also in *Rochester Engineer* 54, no. 10 (May, 1976): 182.

"An Answer Is Blowing in the Wind." *Progressive,* January, 1976, p. 43.

"Community Energy Systems." *Bulletin of the Atomic Scientists* 35 (1979): 50.

"Minimum Deterrence, Maximum Stability" or "Disarmed Deterrence." *Bulletin of the Atomic Scientists* 41 (1985): 42.

Chapters in Books

America Armed, edited by R. Goldwin. Chicago: Rand McNally, 1961.

Nuclear Weapons and the Conflict of Conscience, edited by John C. Bennett. New York: Scribner's, 1962.

Problems in World Disarmament, edited by Charles A. Barker. New York: Houghton-Mifflin, 1963.

The Atomic Age, edited by Morton Grodzins and Eugene Rabinowitch. New York: Basic Books, 1963.

Unless Peace Comes, edited by Nigel Calder. London: Penguin Press, 1968.

World Topics Year Book 1968. Lake Bluff, Illinois: Tangley Oakes, 1968.